WITHDRAWN

6/06

High Noon in the Automotive Industry

Helmut Becker

High Noon in the Automotive Industry

With 86 Figures
and 29 Tables

 Springer

Dr. Helmut Becker
Institute for Economic Analysis
and Communication
Laimer Straße 47
80639 München
E-mail: dr.becker@iwk-muenchen.de

Setting and Layout: Institute for Economic Analysis and Communication

Cataloging-in-Publication Data
Library of Congress Control Number: 2005935913

ISBN-10 3-540-25869-8   Springer Berlin Heidelberg New York
ISBN-13 978-3-540-25869-8   Springer Berlin Heidelberg New York

Springer is a part of Springer Science+Business Media
springeronline.com

© Springer-Verlag Berlin Heidelberg 2006
Printed in Germany

Cover design: design & production GmbH
Production: Helmut Petri
Printing: Strauss Offsetdruck

SPIN 11425083   Printed on acid-free paper – 42/3153 – 5 4 3 2 1 0

Preface

This book was born from curiosity.

To begin with, it was the curiosity of an *economist* who studied in the 60's in an environment which has subsequently developed from *national* into *global economics*. Who has to recognize that politicians, scholars and large segments of society oblivious to supranational authorities and economic globalization forces continue to labour under the notion that they are still fully autonomous and sovereign when shaping national economic policy. And pretend as though their own national state were still the "master in its own house" that despite unbridled market economics could continue to dictate to the economy and companies how to live and in which "rooms".

All that has become fiction. The *laws of globalization* diminish the *manoeuvring space for shaping* national economic policy. Even if many folks today don't want to hear it: The issue is no longer achieving what is socio-politically desirable for the own society but rather the optimal adaptation of society and social benefits to the politically practicable.

By the collapse of communism and the bipolar political world at the latest, the general conditions under which the established industrialized nations could act economically had changed fundamentally. Whereas up to then only the free market-based national economies with just 1/5 of world's population were allowed to take part in the annual competition for the highest GDP growth rates, the title of the "world export champ" or the greatest standard of living increase etc., this pool was abruptly enlarged at the beginning of the 90's to about 4/5 of mankind.

Since then the rules of the game for global competition have been redefined. All of a sudden, economies with huge raw material and labour resources as well as high "unsaturated" market potentials announced their claims to take part in the competition. The process of globalization began. And it developed with breathtaking speed! Barely 15 years have passed since the iron and bamboo curtains fell, and discourse about economic policy in western industrialized countries is already ruled by growing concern over the consequences of this *globalization* for prosperity and jobs, and in Western Europe, particularly Germany, capitalism bashing in turn is becoming ever louder.

The economist faces a series of questions: What happens when a national economy is replaced by the global economy? What happens to a highly developed national economy that suddenly has to deal with competitors in a global marketplace who boast almost equal productivity and quality levels but highly unequal wage levels?

That is one source of curiosity. The other is a by- product of the author's close relationship to the *automotive industry* and its analytical and strategic problems which have occupied the major part of his professional life. Because as it happens this industry represents one of the key sectors, especially in high industrialized countries as Germany, particularly affected by the rapid change in the fundamental structure of the global economy. On the one hand, there is the global re-evaluation of production locations due to significant differences in factor costs, sometimes as close as in their own backyards. On the other hand, there are unmistakable growth limits in their primary markets. The result has been a free-for-all battle over market share!

The automotive industry thus has to bear the double burden of market and production site competition. So the key questions are: What happens in and to an oligopolistically structured industry when the foundation is abolished on which it's post-war growth, high profits and social prestige had rested? How will the oligopolists involved react? Which solutions does competition theory offer for the relentlessly intensifying battle over market share between those involved? As the industrial nation with the highest structural dependence on cars, what employment and social benefit issues does Western Europe have to come to terms with? Does Germany's automotive industry as that on top of the cost-iceberg still has a chance in the long run? Can the German automotive industry survive the global adjustment process?

This book is nevertheless not written from German but from global view. It should bring some clarity to the expected global trends and structural changes in the automotive industry and in the national economies most affected by it. However, it must be conceded that, considering the global economic dynamics and complexity of the material, gaps in these findings will be unavoidable, not because they were overlooked, but because they simply are not yet clear in spring 2005. And the author does not have the powers of Nostradamus!

This refers especially to the future role of China and the Chinese automotive industry. The fact is that within a mere decade China, both as sales market and production location has turned into to the third biggest automobile country in the world (passenger cars + trucks) after the U.S. and Japan. This was of no consequence for the "old" automotive world and thus also for this book, insofar as Chinese automobile manufacturers hadn't been present as exporters in the global market and, beyond its borders, Chinese brands were more condescended to than regarded as competitive. This has changed, too! Recently plans have become public in which national Chinese automobile manufacturers (Chery, Geely, Brilliance, SAIC) in the future want to enter in the competition not just as exporters on the world market but even with their own assembly plants in Europe beginning in 2007. And there's more: Between the printing of the German and English editions of this book, the first "land winds" – off-road vehicles made entirely in China – have swept across Europe's coast, and at breathtaking prices.

For now, all the necessary information is lacking for any reliable estimate of the concrete market effects of China entering the world market. Only one thing can be said for sure: The opening bell has sounded on a new round of predatory global competition! Its initial effects will not be visible and thus accessible to an economic trend analysis until 2010 at the earliest.

This book wishes to banish the illusion that all western manufacturers and suppliers can successfully handle the unavoidable adaptation to the pressures of predatory competition. No matter how "smart", "quiet" or otherwise the "revolution" of the added value chain in automobiles may be, economic efficiency gains from global competition are not for free. A single business cannot alter the predetermined global trend. It can only adapt better than its competitors – or give up.

The same is true without exception for national economies. This book attempts to get their political and social "management personnel" to realize that no highly developed, liberal and free market-oriented economy in the triade-countries can offer pain-free solutions or "panaceas" for the necessary adaptation to such an accumulation of shock-like changes in the international economic environment. In this respect Germany is not standing alone, but is standing in the first row! Therefore the "German case" and the solutions that will be finding here will be of special interest for all other automotive western countries. However, the burden of adaptation can be kept as low as possible by returning to the familiar old virtues of rolling up sleeves and common sense and flexibly and bravely preparing for the inevitable, perhaps even drawing advantages from it. For the *global economy* doesn't just mean risks for the old industrialized countries but also opportunities.

It must be admitted that the results of the forecast through 2015 and the conclusions drawn from it are not pleasant *ad hoc*, either for the automotive industry, economic policy or the individual social groups affected. But it is well known that "Self-awareness is the first step toward improvement." The future for the *old economies* is not to be won by timidly adhering to the past or *in the sleeper car* (German President Horst Köhler) but only with a boldness for global competition and readiness for change. Although Germany has a lot of homemade problems it can nevertheless be taken as an example for the global community. If the German economy hadn't once already summoned up this pioneering spirit after the war, there wouldn't have been any "economic miracle". In other words: fighting famine with diets is no strategy. If this book is able to contribute to a renewed faith in this home-grown strength in all highly developed countries under global competition pressure, the effort will surely have been worthwhile.

Books of this kind have many fathers. I am very obliged to Adolf Ahnefeld and Franz-Josef Wolf for numerous suggestions and above all for the encouragement to finally take a critical and unvarnished look at the future of the global automotive industry from a theoretical perspective of global competition. Likewise many thanks to Werner A. Müller as well as Ruth Milewski of Springer Verlag who made the publication of this book in the present form possible through their kind suggestions on its form and contents.

Special thanks are due to my assistant Juri Dutka for his indefatigable research work, analyses, conception and calculation of the IWK Survival Index *(ISI)*, creation of graphics etc. Also great thanks to Niels Straub for his expert contributions on the components supply industry and the almost daily updates with new reports about labour agreements, location shifts, suppliers takeovers etc.

Finally, I would like to thank my wife for her critical scrutiny of substantial passages of text and for kindly letting me spend so many weekends with my manuscript.

Helmut Becker

Munich, May 2005

Contents

Introduction

"The important thing is not to stop asking"

Albert Einstein

Globalization, market saturation, falling prices, market atomization through endless model offensives by all manufacturers in every imaginable market segment, innovations of sometimes questionable customer benefit, and the increasing technical and organizational complexity of the product automobile characterize the demands of the automotive industry at the beginning of this 21st century.

The battle for markets and clients in the sector has reached an unprecedented ferocity. The consequences are far-reaching structural and regional reorganization along the whole added value chain. For manufacturers, and particularly automobile suppliers, this means clear strategic business plans are necessary to protect their future sustainability. The focus is increasingly on "stay or leave" decisions with regard to present markets and traditional production locations. The latter in particular is hitting Germany.

Exhaustive treatments are legion by scientists and management consultants about

- the coming reorganization of the added value chain in the industry, including projections down to the last cent about future sales-based work division between manufacturer and supplier,

- success factors for manufacturers and suppliers in the future added value chain,

- as well as exact proportional forecasts of technological trends in the product and production process.

Their knowledge normally is based on the customer surveys of consultancies relevant to specific links in the added value chain (manufacturers + suppliers), or it is comes from technology trend updates by scientific and technical research groups.

All these studies are valuable and offer those original equipment manufacturers (OEMs) and suppliers important strategic advice on how to act in order to protect their profitability and competitiveness. All the while they fail to note that in oligopolistic markets like today's automobile market the simultaneous implementation of the measures they suggest is of little use to those involved when the result is in an even more tense competitive situation with ever lower yields. Because not only the consultants but also the measures have their costs.

The discount battles that have been raging in recent years between U.S. manufacturers in the American automobile market are a good example for this: The endogenous growth dynamic of the U.S. automobile market as a whole has been unaffected by the discount campaigns of the various manufacturers, meaning its development has been and will remain predetermined exogenously by the macroeconomic / overall general economic conditions. Instead every U.S. manufacturer reported individual heavy losses on the micro level, making the American consumers glad because their "consumer surplus" rose with the fall in prices. Thus it follows: In narrow oligopolistic markets, if equally strong suppliers are running at the same speed in the same direction, nobody will get ahead! That's just the way economic theory is.

Individual measures by individual manufacturers/makers don't change the market. If a market is showing long-term (structural), non-temporary (cyclical) overcapacities, capacities have to be reduced, meaning makers will have to leave the field – all these business management prescriptions won't help! It may well be correct, say, that the part supply industry's share of global automobile added value in the next 10 years will increase by 12 % or more than US$ 300 billion, corresponding to the current GDP of Switzerland, at the cost of the OEMs. But this totally neglects the fact that, because of increasingly intense global competition for macroeconomic reasons, the number of OEMs and suppliers will nonetheless shrink considerably. The forecasted added value increase clearly won't reach everyone. Therefore the strategic imperative for both manufacturers and suppliers must be to take a stand such that one can cross "the finish line" with the others. And if, like Toyota, the goal is to win, then one has to be particularly good.

Thus, one thing becomes obvious: The question of why are rarely asked in traditional, purely microeconomic consultants' recommendations. What reasons, what general macroeconomic conditions are driving the expected market changes now and in the future? What will happen if the subjective estimations provided by those parties are wrong? Or if exogenous structural breaks cause technological trends to develop differently and the global economic conditions specific to automobiles take an entirely different course than what has been charted in the trend extrapolations by the makers surveyed?

The question about the deeper reasons for the exogenous changes in the competitive constellations of the world market, and above all their consequences for OEM and suppliers, mostly not only isn't answered but generally isn't asked at all. Otherwise statements like "The logical consequence of these circumstances (meaning: market saturation and decreasing sales opportunities in the triad; authors note) is the shifting of competition into other regions".[1] Quite the opposite! The old volume markets are under fire!

This is where this book begins. The core intention of this book is *not* to spread strategies for OEM and suppliers to successfully adapt to changing added value chains. Rather the main issue is *which fundamental forces* are effecting *these changes in the supply (=costs) and demand (= market growth) conditions in the global automotive industry*. First of all one has to ask *why* something is happening, before one can say *which consequences* it will have for the entire industry and the traditional production sites. The reader can then derive his own strategic recommendations. There are plenty of hints about those.

The present study is trying to perform this. The operating procedure of the IWK hereby is "classical": First of all the *diagnosis* is made, thus building up the *prognosis* and last but not least are given strategical recommendations in a general manner for the supplier industry.

In **chapter 1**, the reasons for the drastic intensification of global competition in all automobile markets at the beginnings of the 21st century will be analyzed. How have the conditions in global sales markets changed? Which consequences does that imply for the sector as a whole? What are the symptoms of this worldwide radical change in the industry in the first decade of this century? How are manufacturers themselves responsible for the spreading fall in yields? How affected are suppliers and which ones are particularly affected? What has to be done to "survive"?

This study takes a particularly close look at the German automotive industry in respect of its leading role in Western Europe. In **chapter 2**, internal and external reasons are identified which have been undermining the competitiveness of German automobiles. Here the analyst soon faces the problem that *the German automotive industry* in that abstract form doesn't

---

[1] Kurek, R. (2004), p. 131

exist at all. Instead he has to deal with six automobiles companies[2] with a number of self-managed brands (e.g. AUDI, Bentley, Mini Cooper, Smart, Rolls-Royce, Skoda etc.) and what has become a confusing multitude of model and production series.[3]

How heterogeneous those companies were in 2004 can be seen in their different net returns on sales, which vary between +17% at Porsche to – 4% at GM daughter Opel[4]. Thus, from the start, any statement about the German automotive industry does not proceed from a genuine average. Nevertheless, the present IWK study does not rely on absolute data but on *trends* which in oligopolistic markets are equally valid for all manufacturers, independent of their individual sales and competitive situations. However, it must be admitted that the OEMs with larger yield cushions have more room for reaction and thus more options for dealing with negative trends in a more long-term and forceful way than manufacturers like Opel who are deeply in the red. When "the attic" is already "burning", the question is moot of whether it is better to wait for rain or immediately start putting out the fire with water.

No less interesting is the question of future developments, particularly one issue: Is there any hope that the predatory global competitive situation will relax in the future? Will the market in the future help to solve the earnings problems, or will manufacturers and suppliers have to manage them on their own? **Chapter 3** points out the *megatrends* of global market development as well as the "new division of labour in the global automotive industry" – regionally and structurally – for the next ten years.

The global predatory competition in the automotive industry begins at the top of the added value chain, the manufacturers, and is passed on to all the up- and downstream levels *("chain competition")*. Even if a progressive concentration among OEMs, and in the same way among suppliers, has been visible ever since the very beginnings of the automotive industry at the beginning of the 20th century, the form and ferocity of today's cutthroat competition are new. Remember: in Germany in the 1920s there were over 300 automobile manufacturers. Meanwhile the number of independent OEMs has dwindled to a narrow oligopoly in which even premium producers complain eloquently about the heightened competition. That may be understandable when seen from the narrower point of view of

---

[2] BMW, DaimlerChrysler, Ford, Opel, Porsche, Volkswagen.

[3] According to current press releases VW alone wants to put 20 new models on the market in 2005.

[4] IWK calculations based on press information. Actual result likely to be lower.

management, but from the broader viewpoint of the economist it is seen as proof of market saturation and functioning competition. This is a benefit to the consumer, who of course on the other hand, as an employee in the automobile industry, feels the full negative effects of these otherwise positive aspects.

The question of how this global oligopoly of the remaining 12 motor manufacturers will continue to develop in times of foreseeable heightened competition is of course exciting: which of them will maintain their position, which will run the risk of disappearing from the market as independent OEMs? This question will be discussed in **chapter 4**. To assess the further development of the global automotive industry, the study examines the 11 largest remaining automobile conglomerates in order to give them a sustainability rating. The *IWK-Survival-Index (ISI)* is drawn up using a detailed rating system which has been developed by us. It measures the sustainability of each individual manufacturer and gives them a weighting value by drawing on a number of codes and valuation criteria. It may take the wind out of the sails of any critics if we mention here that the IWK is not a professional rating agency and has at its disposal neither the material nor the personnel resources to examine each OEM meticulously and at high cost. We have restricted ourselves to analysing the available rating results, adding our own knowledge and assessment, and to evaluating everything according to a specially designed pattern, summarizing it all in the *IWK-Survival-Index (ISI)*.

From the perspective which the automobile manufacturers as a whole have on the future one can infer the consequences for the supply industry. *What will happen to the suppliers when the OEM's oligopoly tightens up?* How should they react to the concentration which is to be expected in their clientele, if they want to survive? These are the questions **chapter 5** attempts to answer. Here too, as with the OEMs themselves, a universally applicable recommendation cannot be made. This, although many relevant analyses by the branch's experts would seem to suggest it were possible. Here too, as with the OEMs, there is no one typical or average supplier, quite apart from the differences given by their respective positions in the added value chain. Publishings on this topic are already very extensive. Therefore this account limits itself to the challenges and risks facing the supply industry as a whole in Germany, and subsequently evaluates the deciding strategic success factors for survival in the automobile world of the future. However, here too it is clear from the beginning that not all suppliers will make it to the other side and the number of independent suppliers will shrink considerably. Here it is necessary to build up market power in order to be in a position to meet OEMs as equals and without bribery.

Last, but not least, chapter 6 deals with the future global map of location for the automotive industry. With respect to the high geographic density of automotive activities there and the low cost level in the very near neighbourhood of Eastern Europe countries there will be given a special view on *Germany's future role as a location for the automotive industry*. Does Germany have any chance as a production location in the automotive industry? In this globalized world, in which it has seemingly over night become possible to operate automobile plants in (almost) every part of the world, but at considerably lower cost? If one followed pure economic theory about perfect factor markets, the answer would be simple but painful: as far as the figures go, in the global location competition for the production and assembly of primary and end products, whether labour-intensive or simple, the high-cost location Germany would not be able to keep pace with the low-wage countries of Eastern Europe and Asia. And these reflections are also transferable to other automotive countries in Western Europe.

Indeed, just such a displacement of production and employment has long been going on. The question is:

- How far and how deeply will this location arbitration cut into industrial and employment structures?

- Whether German automobile producers have got through the worst blood-letting of value addition and employment through the agreements already made with the workforce on the safeguarding of the future and of employment? Or are there already first signs that global location competition is penetrating into mass production of up-market and complex products, and indeed even into the development of such products – the core competence of the German automotive industry?

- What arguments point in favour of Germany continuing to be an automobile location, albeit as a "plucked goose", in the long term?

However indispensable entrepreneurial decisions based on the harsh cost and rentability constraints of business management may be for the processes of alignment and restructuring in the German automotive industry, the courage and determination required of all parties should not be underestimated, regardless of how much acceptance this necessity finds within society. The challenge presented by globalization must be accepted *on the offence*, and not *on the defence*. Leadership, decisiveness and creativity are required of managers. Employees are required to act in a manner which shows they understand that it is in their own interest. It is not wishful thinking and illusions, - nor yet lobbyists and functionaries – which make for secure jobs, but competitive products, and if need be, working harder and longer.

In times of universal and general uncertainty about the effects of globalization it is down to the politicians to make sure that economic activity doesn't suffer from home-grown hindrances, but is promoted. Therefore the responsibility for employment, education, financial, and economic policy in particular must lie with the politicians. They must create the necessary framework and get rid of hindrances. One doesn't need to look far!

The requisite analyses and proposals for policy makers in "Old Europe", especially in Germany, are legion and will not be dealt with any further in this book. The author can add nothing new to them. He can at most encourage policy makers to overcome the egotism of groups and functionaries and to act courageously – at the risk of treading on the toes of the sovereign.

1 The current situation: markets in upheaval, turbulence in the sector oligopoly

1.1 Overview

In the global automotive industry the symptoms of a growing, structural profit crisis are increasing. The German automotive industry, up to now a pillar of the German economy, is no exception. This was made more than clear by the spectacular cuts in labour and wage agreements all across the industry in 2004. Even premium makers like Audi or BMW at the beginning of 2005 had to admit to tighter earnings, despite good sales figures in 2004. From which one can conclude that the creeping deterioration of earnings by now has reached the upper market segments.

For several years, in some cases more than a decade, the big global volume markets, whether the U.S., Europe or Japan, have been showing no or only little growth, or are even shrinking, as has been the case now in Germany for four years running since the beginning of 2000. For an industry like automobiles which has been among the motors for growth and income since the end of World War II this absence of growth has been a completely new phenomenon!

This alone would be sufficient in this strongly growing industry with a major social reputation to provoke at first agitation and consternation and then a hectic competitive sanctionism for customers.

But things are getting even worse for the global automotive industry because at the same time, in a breathtaking expansion of model and engine portfolios, each manufacturer is trying to defend its own market segment "with tooth and nail" or to enlarge it at their competitors' expense. The immediate consequence of this is that even one-time niche makers have been turning into "full line makers". The result of these individual portfolio expansions are global production capacities which far exceed worldwide demand and thus are leading to *growing overcapacities*.

Therefore, not only are investments for new production capacity rising progressively for all manufacturers but so especially are costs for development and "market conquering". Simultaneously intensifying price and discount competition is shrinking unit profits. The consequence is brutal predatory competition which is forcing down the profit margins of all automobile producers step by step along the chain – as to be expected in an oligopolistic market with such aggressive pricing. Niche and premium suppliers are also affected, for example Porsche, BMW and DaimlerChrysler. No supplier can evade this cascading competition, sooner or later it will reach every manufacturer.

The consequences are simple: where there is competition the consumer is happy and the producer grieves. Where there is crowding-out it hurts the marginal producers first of all – at least as long as they are still in the game and able to feel the pain. The attempt to defend their market position, for example by means of aggressive discount and pricing policies, and to regain their earning power leads to the corresponding reactions of more efficient suppliers, whose earnings thus also come under pressure. It is a vicious circle from which no manufacturer can escape.

Automobile manufacturers all answer this oligopolistic, unusually heightened competition in text-book manner: cost reduction programs and savings concepts dominate the activities of the automobile conglomerates. These range from intensified pressure on components suppliers and special programs aimed at lowering manpower costs in Germany to closure or the shifting of production into low-cost foreign countries. Thus the most important topic at the moment in the automotive industry worldwide is cost reduction.

Cost reduction programs at the top of the agenda for German automobile manufacturers at present. This applies either publicly (as with the re-capitalization of Opel, and the wage negotiations at DaimlerChrysler and Volkswagen) or closed, and therefore not perceivable, to the public. Examples of this are BMW and Porsche, who don't want to miss out on the cost advantages which their competitors have negotiated with difficulty and with very negative press coverage.

So much must be mentioned in advance at this stage: an efficient cost structure is merely a necessary, but not a sufficient condition for securing earning power. The decisive factor for success in the market is still the attractiveness of the product and not how efficiently it has been produced. Thus Porsche has for years, as a small niche supplier, been earning the highest yields in the entire global automotive industry, even though the company produces almost exclusively in Germany, a high-cost location.

In the past few years it has especially been volume producers in the lower and middle market sectors such as VW and Opel, but also Daimler-Chrysler, who have got into earnings crises, albeit in part for completely differing reasons. Competitive pressure is coming from two sides in the midrange segment which makes up almost three quarters of the German market. Since the end of the 90's the premium producers have been poaching on compact car territory. The Mercedes A-class, the Smart and, as the newest example the BMW 1 series are proofs that this offensive does not stop at the borders between market segments. Big brands can also make sales in the middle and lower segments.

At the same time the Koreans and Japanese have been threatening the traditional market segments from below, making life doubly hard for German producers. The Korean manufacturers Daewoo and Hyundai show double-digit growth rates in the European market, where they are successfully expanding their position after starting at low level. The same goes for Toyota and Mazda, who are gaining market share year for year. The Asian brands offer extremely reliable, solid models with a good price performance ratio. Starting from this base they are now slowly but surely beginning to advance into the premium sector (for example, the Toyota Lexus).

It's the same picture in the USA. Here Japanese, Korean and German makers are successively crowding out the remaining two American makers, GM and Ford. Significantly, both firms have been continuously operating in the red; profits are generated solely by financial services, which is really a core business of financial establishments (banks) and not of manufacturing companies.

To summarize, it can be said that in early 2005 the worldwide automotive industry is in a state of oligopolistic destructive competition. As the market as a whole is no longer growing, every producer is trying to generate growth at the cost of the other competitors. It is obvious that these aren't ready to give in without putting up a fight. The result in the end is that none can be the lucky winner; to a greater or lesser degree they are all losers with stagnant market volumes and shrinking profit margins.

The branch is caught in a dilemma caused by slack consumption, advancing Asian competition, growing overcapacity and costs due to growing model diversity, tougher price competition, and passing losses on to components suppliers.

1.2   Structural growth weakness in the triad

The automotive industry in the triad (The USA, Western Europe and Japan), which represents more than two thirds of the world's automobile sales volume, finds itself at the beginning of the 21st century in a situation of *distinctly slack demand.* This cannot be explained merely by normal cyclical economic weakness after a preceding economic high. In Japan the number of new registrations after almost 15 years of continuous market weakness is still below the figures for 1990 (5.1 million passenger vehicles) and shows only a limited recovery. In the USA sales of passenger vehicles (including light duty) are receding for the fourth consecutive year. The number of new registrations in Western Europe has also been below the values of the late 90's for five years now.

Fig. 1. Development of new PC registrations in the triad

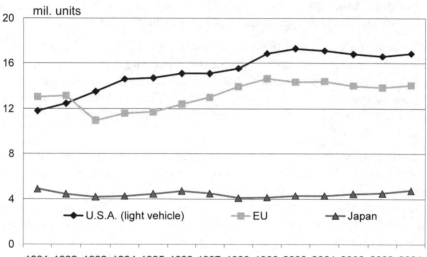

source: VDA, IWK presentation

In the same way, *Germany,* Europe's largest volume market, shows a persistent sales volume weakness. Since 1999 new registrations of passenger cars in Germany have sunk by 15% and are now stagnating for the third year in a row at a level below 3.3 million. German car makers are especially affected by this long period of sales weakness in their home market because of their market share of around 70%, and with the exception of a few (Porsche, BMW) are presently suffering from considerable yield problems. This has already lead to serious structural adjustment such as the revision of wage settlements, withdrawal of shift allowances, comprehensive programs to reduce fixed costs, and in some cases massive reductions in the workforce. Even DaimlerChrysler's premium brand Mercedes brought them their worst fall in profits for a long time in 2004 (-47%).

Recognition is growing worldwide that growth in the form which had been considered usual until recently is now a thing of the past for automobile manufacturers. Because of the saturation of traditional markets, global competition for market share has become distinctly more intense and competitive pressure has increased drastically for all involved.

Fig. 2. Development of new PC registrations in Germany

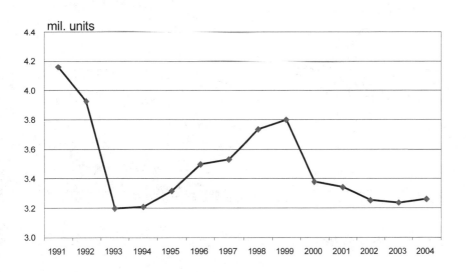

source: VDA, IWK presentation

In this form, this is a new for the automotive industry. The market for motor vehicles used to be determined far less by demand than it is today. In the period from the end of World War II until the end of the 70's the German automobile market had all the features of a classical seller's market. Customer taste and quality needs varied only slightly, due to the high level of unsatisfied basic demand. Waiting times for the delivery of vehicles sometimes amounted to as much as 4 years (Daimler Benz).

In view of the undamped growth in the demand for German cars and the relatively low level of competition from import vehicles still prevalent at that time, there was no strategic necessity for German automobile manufacturers to adapt to the special needs of smaller demand groups. Export was an additional reliable driving force of growth, as German OEMs were positioning themselves excellently with innovative products and good brand identity / image, which they have been able to expand continuously until now.

Fig. 3. German car exports

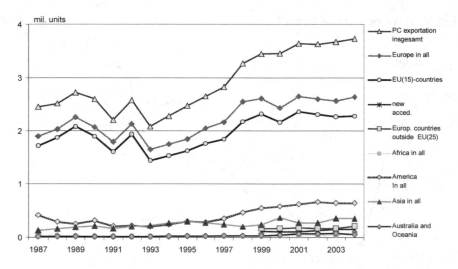

source: VDA, IWK presentation

This success in the important export markets of the USA and Europe enabled German manufacturers to initially face effectively the weak demand in the domestic market, the growing competition from a French market which had regained its strength, but above all, the massive competition from the Japanese and Korean automobile industries since the beginning of the year 2000. In particular, those German manufacturers were successful, who were exporting to countries outside Europe. Those OEMs, however, Ford, Opel, and Volkswagen, for example, who focussed on

Europe, found themselves increasingly under pressure, even though it was the French and Italian brands that were the biggest losers in the face of the competition from Asia. Even though German manufacturers had to accept losses in the domestic market due to the increased competition, they were still able to increase their market share in Europe slightly to 46% in 2004 despite harsher market conditions.

1.3   Competition from Asia intensifies

In 1960 the share of foreign automobile brands in new registrations in Germany stood at 9.7%, but then rose continuously to 32 – 34% by the beginning of the 90's and reached about 36% for the first time by 2004. It was mainly the Asian manufacturers – first the Japanese, but meanwhile the Koreans, too, who were able to achieve a significant increase in their market presence in the whole of Western Europe, and to take market share from the established domestic manufacturers.

European countries with their own automobile industries (Germany, France, Italy, Great Britain, and Spain) reacted in the early 1980's to the Japanese manufacturers' successful market entry by initially erecting import barriers for Japanese cars. In the context of the European domestic market, these national regulations were replaced by a transitional regulation ("Elements of Consensus"), which restricted Japanese vehicle imports to a 15% share of the market right up to the year 2000.

However, this quota was never exploited. On the contrary: against all expectations exports of Japanese cars to the EU did not rise when this restriction ended. Instead, the European market was conquered from Japanese manufacturing plants in Europe, which had been set up in recent decades simply because they weren't forbidden. Local production showed strong growth, whilst the number of imported Japanese vehicles decreased distinctly, analogous to the development in the US market.

While Japanese manufacturers have been able to record continuous growth in the USA from American / Canadian production and have gained a market share of almost 30% there in the meantime, they have had to struggle with difficult market conditions in Europe. In contrast to the US market, which is dominated by the "Big Three" (GM, Ford, Chrysler), with its "value for money" mentality, The European market is essentially more heterogeneous, with far more competitors – and above all with more demanding customer expectation the sales volume of Japanese cars in Western Europe fell significantly during the 1990's and amounted in 2001, for example, to about 1.5 million units, which was still below the level for 1990. However, the situation changed drastically in the following years. Lead by Toyota, new registrations of Toyota vehicles in Europe recorded a considerable rise, despite a generally weak market. With annual growth rates as high as 10% (2003) they achieved a record value of 1.9 million cars in 2004.

This market success of the last three years is all the more remarkable because of the decline in the market as a whole during this time, and because of a distinct drop in sales which European mass producers had to some extent to accept. Only the Korean producers were able, with over 20%, to achieve more growth than the Japanese in the last two years, albeit at a very much lower level of fewer than half a million cars sold.

Fig. 4. PC-new registrations in West Europe

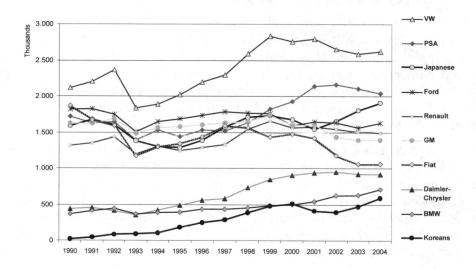

source: ACEA, IWK presentation

The advance of Asian producers into the European market is more clearly recognizable in the development of market share. In the 1990's the share of Japanese cars in new registrations remained relatively constant between 11% and 12%, thus lower than would have been possible according to the trade restriction agreement with the European Automobile Manufacturers Association (ACEA). Since 2001 this share has been rising continuously to over 13% by the end of 2004 (Fig. 5).

In comparison to the USA, where Japanese brands' market share is twice as high as in Europe, the value on the European market still looks relatively small and obviously allows scope for increase. It is a well-known fact that Toyota wants to continue in its aggressive treatment of the European market, increasingly at the top end of the luxury class. "We will continue our strategy of developing, designing and constructing cars for Europeans in Europe". (Akihiko Saito, Toyota's Executive Vice President and R&D Manager). [5] For this reason Toyota wants to invest around 75 million Euros in the R&D center at Zaventem (Belgium) by 2009.

Fig. 5. Development of market shares in West Europe

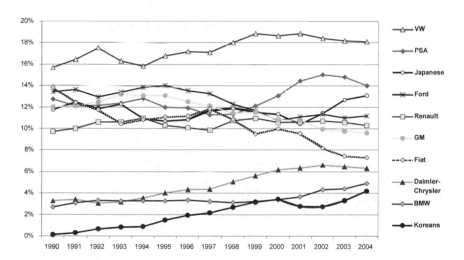

source: ACEA, IWK presentation

The Korean manufacturers, in Europe only since the beginning of the 90's, were able to move ahead no less sweepingly. Further market penetration is to be expected due to better quality and considerable progress in the Europeanization of their style. Their market share stands at over 4% at present. Tendency: increasing (Fig. 5).

---

[5] Automobil-Produktion (2004-04-07)

For the strategic evaluation of the Japanese conquest of the market it is worth observing that the strong growth in Europe was not shown to the same extent by all the Japanese OEMs, but rather was triggered almost exclusively by Toyota (Fig. 7.). Until 1998 the largest Japanese producer took only 2nd place on the European market, behind Nissan. It wasn't until the second half of the 90's that Toyota greatly expanded its European business and was able to increase new registrations from about 300,000 cars (1995) to about 725,000 cars (2004), almost 150% (Fig. 6.). Thus Toyota was able to gain a share of almost 5% in the Western European automobile market, more than BMW, for example. Consequently, Toyota now heads the table by far among the Japanese OEMs in Europe and wants to grow further, as it has stated. For a targeted global market share of 15%, broken down to fit Europe, this means a further increase in volume to the scale of Fiat's whole sales volume.

Fig. 6. Market shares of Japanese OEMs in Western Europe

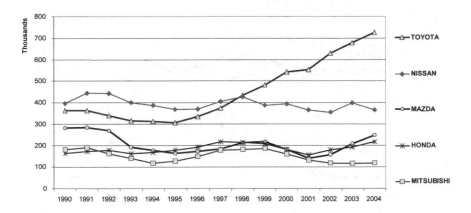

source: ACEA, IWK presentation

In contrast, the other Japanese manufacturers showed no essential growth in sales (Fig. 6). The reasons for this may be varied. There may be too few convincing models offered, or it may simply be from lack of interest or insufficient distribution structures. But it may also just reflect the positions of the European parent companies, who are afraid of in-house-competition.

One example of this is Nissan. Having been restructured and dominated by Renault since 1999 this Japanese OEM has remained fairly constant, with sales of about 400,000 cars and thus takes 2nd place among the Japanese manufacturers in Europe, with a market share of just below 3% (Fig. 7).

Fig. 7. Market shares of Japanese OEMs in Western Europe

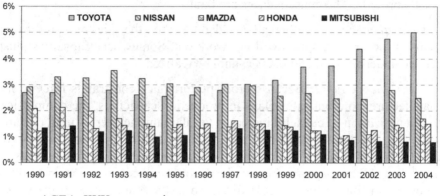

source: ACEA, IWK presentation

The European car sales volume of Mazda, which has been a 33% subsidiary of the Ford conglomerate since 1996, sank by about half between 1990 and 2001, from about 300,000 to a mere 150,000. Since then, obviously spurred by Toyota's success, an about-turn in strategy has become visible, resulting in appreciable growth, so that Mazda has since moved back to 3rd place among the Japanese firms in Europe, with a market share of 1.5%. In order not to put the successful stabilization of its own brand at renewed risk, however, Ford may well have little interest in Mazda making further gains in market share.

In comparison to the US market, the European market did not play an overriding role for Honda and Mitsubishi. Both manufacturers hardly increased their sales of passenger cars in Europe in the last 15 years and were relatively constant at fewer than 200,000 units p.a. sold. However, Honda, too, seems now to have discovered Europe and has shown a slight upwards tendency since 2002. In contrast Mitsubishi's downward slide continues; after the separation from DaimlerChrysler it is only a matter of time before it disappears from the European market (and elsewhere).

1.4   Growing overcapacities, decreasing capacity utilization

Despite slack sales worldwide in the largest volume markets, the manufacturers' production capacity continues to grow unabated. Apart from "normal" increases in capacity driven by productivity, the reason for this paradox development is quite simple: The lemming-like investment behaviour of all manufacturers in extending their model ranges simply cannot harmonize with the given exogenous market volume. To put it another way: Everyone is planning to take a bigger piece of the cake than the cake can finally provide. The consequences are fatal!

Consequence no. 1: When all is added together, the sales and production plans of the manufacturers will not work out. Somewhere capacities must remain under-utilized and plans cannot be realized.

Consequence no. 2: Because profit performance constraints force every manufacturer to align sales plans to production capacities, the compulsion to full utilization of capacity increases for all manufacturers – until at some stage the warehouses overflow with vehicles and distribution boards request production cuts. Any lowering of peak production inevitably leads to a smaller utilization rate, higher fixed costs and sinking earnings. This would seem true to the old rule of thumb in the automotive industry: *the most expensive car is the one which isn't built!*

Fig. 8. "Unplanned" overcapacities 2009 (schematical description)

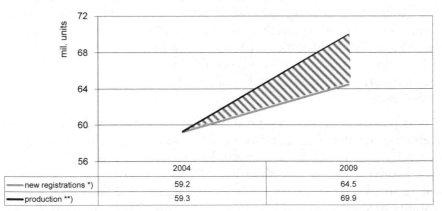

	2004	2009
new registrations *)	59.2	64.5
production **)	59.3	69.9

*) world-wide, valuation IWK. hypothesis: Decline share of triad from 66 per cent to 62 per cent;

**) source: CSM world-wide: apportionment by manufacturers see appendix 3

Table 1. Completion, production capacities and capacity utilization of the biggest automobile groups of affiliated companies in 2003

	completion	capacity	difference	capacity utilization, in %
General Motors Corporation	12.114.941	16.399.373	4.284.432	73.87
Ford Motor Company	7.662.682	10.852.009	3.189.327	70.61
Toyota Motor Corporation	6.726.273	7.914.524	1.188.251	84.99
Nissan Motor Co., Ltd.	5.483.899	7.434.650	1.950.751	73.76
Renault S.A.	5.483.899	7.434.650	1.950.751	73.76
DaimlerChrysler Plc.	5.443.318	7.216.078	1.772.760	75.43
Volkswagen Plc.	5.118.449	6.756.239	1.637.790	75.76
Peugeot S.A.	3.318.224	4.056.254	738.030	81.81
Honda Motor Co., Ltd.	2.943.854	3.443.785	499.931	85.48
Fiat S.p.A.	1.183.736	1.827.966	644.230	64.76
BMW Plc.	944.072	983.408	39.336	96.00
Total	**56.423.347**	**74.318.936**	**17.895.589**	**77.84**

source: PWC, IWK calculation

The matter of overcapacities per se is not new in the automotive industry. It could almost be said that the automotive industry has learned to live with them since the sales slumped as a result of the oil crisis in 1973 and the following years, and to take them into account in their profitability calculations. The fact is that the breakeven point for all manufacturers is today much lower than it used to be. How necessary that is can be shown if on takes Smart as an example: in 2004 a loss of 600 million EUR was made, at 50% capacity utilization. The Volkswagen group is aiming at a breakeven point at 65% capacity utilization. At the moment it ranges between 70% and 72%.

A reduction of global overcapacity in the automotive industry cannot be reckoned with in future. Experts' projections assume that capacities will exceed the amount of actually produced vehicles by about 20 million units per year in the next few years.

Fig. 9. Entire production and capacities until 2011

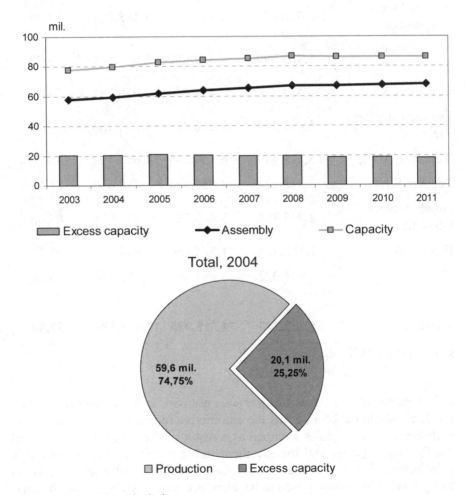

source: PWC, IWK calculation

As Table 1 and Fig. 9 clearly show, there is no doubt that very high overcapacities exist in the automotive industry (approx. 17-20 million units), but there are fundamental differences in their evaluation. Is it a case of overcapacity when a change is made from a 3-shift mode to a 2-shift mode because of an economic downswing in demand? Or when an assembly line is no longer fully utilized because a vehicle model is being phased out?

To be precise, this is simply a case of an *economic or planned operational underutilization of production capacity*, which can be bridged within a short time by means of working time schemes such as short-time working or reduced by the introduction of new models. Vehicle production potential is in this case only temporarily underutilized, not permanently.

Structural or permanent overcapacity is a different kettle of fish. If there is a fundamental lack of sales opportunity simply because a model isn't competitive, factories must be shut down permanently. That costs money, but it is still cheaper than carrying on under capacity. The American automobile manufacturers have become very skilled in this field, due to the pressure of competition from the Japanese in the past 25 years. This experience is now of benefit to them in Europe.

It must also be pointed out that overcapacity in the automotive industry is not exclusively a question of competition and the market as such, but is often a result of strategic misjudgement, i.e. misdirected corporate investment. The Smart is an example of this, having failed by far to meet the sales figures expected by the management. The structural underutilization of production capacity caused by this lead to losses for the DaimlerChrysler conglomerate totalling 2.5 billion Euros – despite a high level of public assistance in the overall investment.[6]

Although the automotive industry is currently in a cyclical slack period, large amounts of overcapacity are of a structural nature. It is true that demand can increase strongly in Germany and other important volume markets over a lengthy period. The automotive industry is hoping to recover as a result of the replacement requirements which have accumulated in the meantime, but this is not a fundamental trend reversal. The future tendency will be limited market growth. Market saturation is too great for utilization of growing capacities to become permanently better in the future.

---

[6] The former DaimlerChrysler manager Wolfgang Bernhard says Mercedes Benz is a "reconstruction case", and "blood must flow". Cited in Spiegel (2005-01-11), p. 90.

In Western Europe capacity utilization in the automotive industry is currently estimated to be approximately 80%. On closer inspection however, considerable differences between the various producers become apparent. According to a survey by WestLB[7], the figures for European plants range from 66% (Fiat) and 96% (BMW). The development of capacity utilization in Western Europe between 1997 and 2006 is shown is illustrated in Fig. 10. (below), based on WestLB forecasts.

Fig. 10. Development of capacity utilization in Western Europe 1997-2006

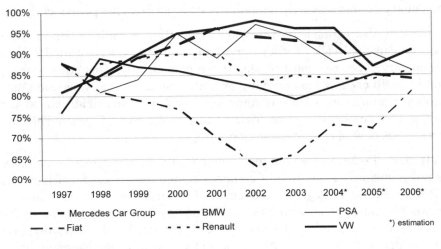

source: WestLB

By comparison, the overcapacities of the US American automobile manufacturers turn out substantially higher because the number of vehicles produced there has sunk in the last few years in spite of an increase in productivity – from a record high of 13 million (1999) to only 12.2 million (2004).

---

[7] Conf. WesternLB (2004b)

Fig. 11. Regional Structure of Overcapacities in 2004

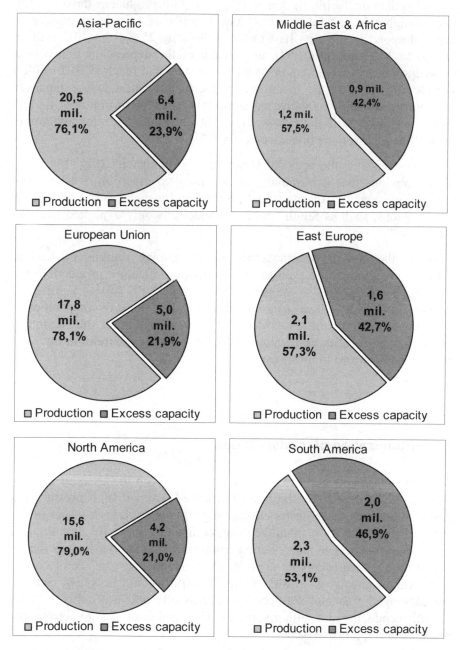

source: PWC, IWK calculations

While American producers still have to react to this situation by shutting down production facilities, Japanese automobile producers have already got through this difficult process. The number of vehicles produced in Japan dropped in the first half of the 1990's by 25% from 13.5 million (1990) to 10.2 million (1994), as a result of the depression. Since then production in Japan has remained fairly constant and reached 10.7 million vehicles in 2004. In the meantime, capacity utilization in the Japanese automotive industry has increased again due to restructuring, lay-offs and factory closures. With the exception of Mitsubishi the other manufacturers have all now passed the breakeven point.

If one considers the regional structure of overcapacity (Fig. 11) it becomes obvious that it is mainly in the so-called emerging markets that production capacity exceeds actual production by such a large margin. In some regions, such as South America, capacity is only exploited to about 50%.

While the current overcapacities in the growth regions of Eastern Europe and Asia can be seen as an anticipation of future production, the growing overcapacities in Western Europe and North America present a structural problem in need of a solution. It is just these overcapacities in the saturated markets of the triad which are responsible for the erosion of manufacturers' profits, and which are likely to grow even more in future as production is moved to low-wage regions.

1.5   Atomization of model ranges

The production and sale of attractive cars have been the basis of all successful OEMs ever since the branch emerged. Even if high profits from financial services make up for shrinking profits or even high losses in operative business in most manufacturing companies, they still do not count as the core business of a production enterprise. In a similar way to that in which, in a macro-economic context, the "modern" service sector cannot exist without the "real" world of the industrial sector, no OEM can sell vehicle-linked financial services so successfully and profitably without actually producing and selling vehicles. This applies at least to automobile manufacturers, if not to banks.

For this reason all OEMs are doing their best to meet the growing crowding-out competition with *innovative products*. In order to meet the ever more sophisticated wishes of their customers and at the same time maintain or expand market share vis-à-vis the competition, all automobile manufacturers have begun to leave their traditional niches or market segments and expand their product ranges "upstream", "downstream", and "cross over". Most important however, are improvements in quality and image. A consequence of this policy, which is very understandable from the management point of view, is that for one thing life expectancy of the vehicles is increasing, while at the same time the growing market fragmentation is leading to a regular explosion of variants and types (Fig. 12).

Fig. 12. More segments and niches

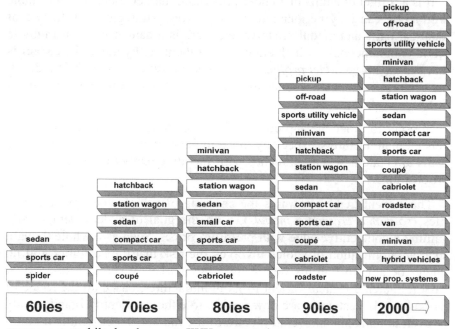

source: automobile development, IWK presentation

The curse of this variety is growing pressure on profits. Every structural expansion of the model range leads to additional costs across the entire added value chain, from development to logistics, production, distribution, and even to recycling. Upward pressure on costs and the compulsion to fully utilize plan capacities for new models do decrease in step with the number of identical parts (platform or modules) used for them.[8] This does not however reduce the necessity to "earn" the additional costs of the new models by producing larger numbers of items. Every new model and every new variant of a model certainly first multiplies costs, and not earnings. Genuine profitability can only follow, if the strategy works out right and market share increases as planned – and not the in-house cannibalism rate.

Nothing will change this home-grown profit and growth pressure until all substantial OEMs tire of announcing "new model offensives unprecedented in the company's history" at every exhibition stand and every automobile fair. [9] Thus the signal given to their competitors is that they must do the same, like it or not, ensuring a further heightening of competition.

It is a further dilemma that under prevailing market conditions profitable production can only be guaranteed by platform strategies and the use of identical parts and modules. However, there is a danger here for multiple brand conglomerates – which almost all of them are by now. The danger is that *brands and product profiles may become diluted*, especially if individual - almost identical - brands are produced at locations with extremely differing costs.

In order to evade the appreciably fiercer competition in traditional markets all manufacturers, especially those in the premium segment, are taking pains to create new market niches with lucrative temporary unique selling propositions.

The current trend is towards the crossing of vehicle concepts and profiles (*so-called crossover models*), with the goal of at least temporarily gaining a competitive edge with premium prices. Seen as a whole though, such an atomization of choice involves the risk that the customer will no longer be able to find his way around the market. Whereas it used to be easy to match a vehicle variant to its manufacturer, based on the shape of its bodywork, customers are now unable to either to identify brands easily or even to maintain an overall picture of the market. To bring it to a point: if niche products are side-by-side the idea of a niche becomes itself absurd, because it is then no longer recognizable as such.

---

[8] Example of BMW: the new 1-series consists to about 60 per cent of parts from the 3-series. A-class and Smart from Mercedes Benz, however, are singular developments with no recognizable identical part procurement.

[9] According to current status quo (April 2005), in 2006 alone nine new cabriolet models will enter the market (from VW, Ford, Opel, Alfa and BMW).

Thus it is all the more important for success to sharpen brand profiles in such a way that the customer feels he is being led safely through this labyrinth, which is the market, by "his brand".

1.6   Change in the buying patterns of "automobile consumers"

Since the beginning of the new millennium car buyers in the European market have shown a distinct change in demand behaviour. So far, not all manufacturers have been able to adapt, as shown by certain spectacular drops in performance and painful restructuring measures. For many decades the motor industry showed a pyramid-shaped demand structure and therefore a pyramidal market structure, corresponding to the classic pattern of income distribution, with relatively few customers in the upper price range, a lot in the middle and most of them in the lower segments.

In the meantime this has changed permanently. Market structure has now assumed the shape of an hour-glass, broad at the top, compressed in the middle, and broader again at the bottom. This is not the shape of a pyramid. Rather, it corresponds to the distribution of wealth in today's society, as has been known to us from the USA for a long time. The so-called middle income group is becoming eroded and is slowly but surely drifting either upwards or downwards in terms of income and wealth.

Demand for vehicles shows a corresponding pattern of development. On the one hand, the premium segment, the top end of the middle range, and the fun segment are favoured (Convertible, SUV[10], Off-road, and on the other hand it is the top end of the lower range which is favoured. But it is the traditional volume segments in the middle range which are in a real fix, previously the domain of the solid middle class. This is precisely the group of buyers that was hit particularly hard by the economic lull which has affected incomes since 2000.

The automobile manufacturers especially hit by this demand structure are therefore the classic mass producers such as Volkswagen, Ford, Opel and Fiat. However, the successes of Peugeot and Renault, who compete in the same segment, show that companies can cope with this changed buying

---

[10] SUV = Sports Utility Vehicle, e.g. Mercedes M-class, VW-Touareg, etc.

behaviour if they control their costs and can offer innovative products and product solutions. Just as the French producers in the field of diesel filter technology, even if there is a large gap between appearance and reality in the view of the German competition.

In short, the sharp drop in the profits of German producers cannot be blamed on the adverse external market conditions alone, but are above all home-made and the result of inappropriate models policy and bad cost management. Thus Opel missed the trend to diesel motors and fun vehicles by years, and Volkswagen missed the off-road and people-carrier trends.

Prices are not about to relieve business results. So OEMs and suppliers alike face the additional problem that, according to the experts, automobile prices will stay at today's level for the next ten years, due to inflation.[11] Stagnant markets, cautious consumers and the heightened competitive pressure among producers are showing effect. The basic model of the VW Golf may be cited as an example of this. Its price has risen only 0.7% since 1990, due to inflation, although the basic equipment was improved and upgraded significantly in the same period of time. The basic version of the new Passat, introduced in February 2005, is being offered at exactly the same price as its predecessor. The competition cannot ignore that kind of signal.

The prevailing crowding-out competition can be seen in the *discount campaigns* used by all the producers in the market. Producers allow their customers appreciable discounts in order to maintain their market shares. Price rebates have been at an average of about 15% for a long time, according to Center Automobil Research (CAR). In the form of combined manufacturer and trader discounts this level is undercut in many cases. In the case of the very severely shrinking premium range, which was considered for years to be particularly discount-resistant, rebates of as much as 30% of the listed price are given, or even more, for "new" used cars. However much the VDA (Association of German Automobile Manufacturers, translator's note) castigates the high discounts in Germany as a blind alley[12] from the marketing point of view, this will not stop the destructive competition.

---

[11] Conf. HAWK-2015 (2003), p. 12.

[12] VDA-president Prof. Dr. Bernd Gottschalk, VDA (2005), p. 2

The fight to win customers by offering discounts is raging particularly fiercely in Germany, as a key market. The following is a historical documentation of campaigns which took place around the end of 2004 and the beginning of 2005. It does not claim to be complete.

The French producer, Renault, whose new registrations in Germany in 2004 had fallen below the levels for the previous year, started the "Rackerwochen" campaign ("rascal weeks"). Car buyers were given a discount on all models, depending on the number of children they had. The discount ranged from 10% (one child) to 16% (four children) – for a car costing 25,000 Euros that's a good 4,000 Euros. Following this campaign, Renault introduced the "commuter weeks" campaign, with similar material incentives. At the end of 2004 Opel offered a price advantage of up to 5,000 Euros on the purchase of special editions. The discount is made up from a combination of free extra equipment and special finance offers. Private customers are increasingly lured into showrooms by cheap finance. Volkswagen is setting its hopes on several instruments, which also add up to considerable discounts. The company is offering its customers cheap finance, cheaper special editions and even a bonus of 1,000 Euros for people who have recently passed their driving test. The Italian competitor, Fiat, is granting higher amounts than the vehicles are really worth on used cars taken in payment. Peugeot and Citroen are also offering great reductions even on new models, such as the Peugeot 307, calling them the "Peugeot Christmas bonus". The discount war doesn't even stop any more for absolutely new models. Prospective customers were offered a discount of 1,200 Euro on Citroen's C4, which is just arriving at garages, even before it came out. The Korean brand, Daewoo, is even offering a banker's check for 500 Euros on purchase of a Daewoo in its Christmas campaign. Daewoo / Chevrolet's latest campaign offers 1% discount per 10cm body height!

The discount war started in the American market, where sales in the automobile market dropped almost completely away temporarily after September 11 2001. In July 2004 alone sales incentives offered by the three biggest American producers averaged at 4,088 dollars per vehicle. In the meantime, the first signs of a relaxation are beginning to show.

On the other hand, customer expectations as regards equipment, comfort (for example air-conditioning), safety, and to a certain extent innovative technology (ABS, ESP, etc.) continue to grow – but not the readiness to pay higher prices for these things. Without doubt, the customer's attitude while making his decision under pressure from economic constraints (narrow budgets and growing employment risks) has become more and more rational in the last few years. The way the general economic framework is expected by the IWK to develop in the next few years, a change in customer attitude – wanting more car for the same money – cannot be reckoned with.

It is easier for premium manufacturers to react to these customer expectations, as they already offer the comfort and image the customer wants but doesn't want to pay for. Mass producers react in their own way, by attempting to meet their clientele's value-for-money fixation with relatively high-quality low-price offers. Thus Renault plans to put the Dacia Logan – a car produced in Romania – on the German market in 2005 for 7,500 Euro. VW is countering with the Fox, produced in Brazil; others will follow.

A prerequisite of such increased offers in low-cost segments is not only considerably lower production costs for the manufacturers, but also sinking buying in prices.

1.7   Intensified overall price and cost competition

1.7.1   The situation of the branch as a whole

Since the beginning of the year 2000 the global automotive industry has seen a considerable drop in yield. Starting from the OEMs, this led to considerable pressure on the margins of their component suppliers. There are many reasons for this, which should be split into two different categories, external and internal, and evaluated accordingly.

With the exception of China and some submarkets in Eastern Europe and Asia, most Automobile manufacturers have seen themselves confronted with stagnating or falling sales figures. Stock exchange crises, weak capital markets, political insecurity, stagnating real incomes, and increasing job risk have forced down the demand for cars and have partly led to fierce price/discount between manufacturers. The consequence is a widespread decline in results. Moreover, "king customer" can sense his market power as a buyer and has begun to demand considerable discounts and other benefits when buying cars. Increasing legal and regulatory requirements serve to further increase the pressure on automobile producers.

One source of external pressure on German manufacturers is that the existing cost structure is still pushing profit margins down hard, despite considerable efforts to rationalize in recent years, which have brought a certain amount of success. At the same time many small and medium-sized components suppliers are paying the price of higher growth-induced structural costs for an increasing level of value creation responsibility, advancement in the supply hierarchy and automatic company growth. [13]

The reasons mentioned have led to a distinct intensification of competition in all international automobile markets. With no quantity expansion any more, the pressure on the direct expenses and overhead costs of the firms – OEMs and suppliers alike – inevitably rose drastically. All links in the automobile added value chain face the situation of having to lower costs radically and permanently.

Feasible reactions for companies are harsh internal cost management in all direct cost/function areas, but above all in the overhead costs; the reduction of capital costs when and wherever possible; increasing purchasing efficiency (which, as we know, accounts for more than half the value added by OEMs) and last but not least, the shifting of production to new manufacturing locations in low-cost countries.

Only the conception and consistent realization of such internal cost reduction and efficiency enhancement measures will enable companies in the automotive industry to survive in the market. Those who cannot will have to leave the marketplace.

New forms of cooperation in the automotive industry, such as co-ops and innovation networks, are a clear sign of intensified competition, where the companies involved form alliances for specific purposes in order to survive. The shifting of production to low-wage countries pushes the trend to increased crowding-out even further. One good example of this is the joint compact car project of the French automobile concern, Peugeot-Citroen (PSA) and its Japanese competitor Toyota. The car was developed jointly by both manufacturers and is being built in three different versions, as the Peugeot 104, the Toyota Aygo and the Citroen C1 in a joint location in Kolin in the Czech Republic.

---

[13] See chapter 5: Consequences for the components supplying industry

This kind of cooperation between manufacturers is not new. The special thing here though, is that it is happening between two companies which otherwise see themselves as direct competitors in the market. It remains to be seen whether such a model represents the start of a development which would take the heat out of the competition in this narrow oligopoly.

This heightening of competition increases cost pressure for the entire automotive industry. Manufacturers in the premium sector with their larger ability to pass on costs are not as affected by this, at least for the time being, but not permanently. Mass manufacturers, however, are the most directly and worst hit. Every single automobile producer must decide for itself whether to do without new innovations, which are only partly wanted by end customers and therefore risky, and which would possibly lead to competitive drawbacks, loss of market share, shrinking yield, and sinking shareholder-value with the corresponding results for employment. Or whether to attempt to economize in other areas in order to compensate for the higher costs induced by new technology.

1.7.2 Profit pressure in core business, but not for all producers

As already shown at the beginning of this study, the automotive industry as such no longer exists, in view of the small number of independent automobile companies, but rather single automobile conglomerates with very different company economies.

How differently the individual OEMs have mastered the structural changes in the branch and the challenges of falling sales and globalization of competition in recent years - despite all the possible methods of book depreciation - can be understood by looking at the developments in key financial figures.

According to the opinion of many financial experts, net profit margin, measured as a proportion of after tax sales profit, gives the best information about the economic health of a company (see Table 2 and Fig. 13)

Table 2. Net sales volume yield of the biggest automobile groups, in per cent

	1994	1995	1996	1997	1998	1999	2000	2001	2002	2003
Nissan	-2.8	-1.5	1.2	-0.2	-0.4	-11.4	5.4	6.0	7.3	6.8
Toyota	1.6	2.4	3.2	3.9	3.5	3.8	5.1	3.9	4.8	6.7
Renault	-	1.2	-2.9	2.6	3.6	1.4	2.7	2.9	5.4	6.6
Honda	1.6	1.7	4.2	4.3	4.9	4.3	3.6	4.9	5.4	5.7
BMW	1.6	1.5	1.6	2.1	1.4	-7.2	2.9	4.9	4.8	4.7
PSA	1.9	-	0.4	-1.5	1.4	1.9	3.0	3.3	3.1	2.8
GM	-	4.2	3.1	4.0	1.9	3.3	2.4	0.3	0.9	1.5
VW	0.2	0.4	0.7	1.2	1.7	1.1	2.4	3.2	3.0	1.3
Ford	-	3.0	3.0	4.5	4.2	4.5	3.2	-3.4	0.2	0.6
DCX[14]	3.8	-1.5	4.2	5.6	3.7	3.8	4.9	-0.4	3.2	0.3
Fiat	1.5	2.9	3.1	2.7	1.4	0.7	1.2	-0.8	-7.1	-3.9

source: annual reports, *ComStock,* IWK calculations

Fig. 13. Net sales volume yield of the biggest automobile groups, in per cent

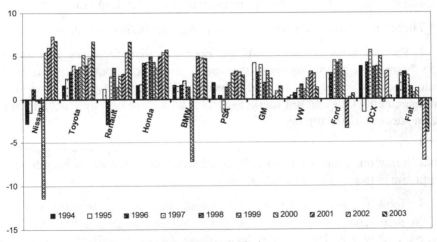

source: annual reports, *ComStock,* IWK calculations

---

[14] DaimlerChrysler

If one follows the development of the net profit margins of the 11 biggest automobile companies in the last 10 years, one can notice a tell-tale peculiarity: there is a group of "winners" with high or even rising profits (Toyota, Nissan, Honda, Renault, BMW), and a group of "losers" with low or falling profits (GM, Ford, DaimlerChrysler, PSA, Volkswagen, Fiat).

By averaging, for the sake of simplicity, the historical net profit margins of the companies, and then calculating from that the trend for each group (Fig. 14) a distinct split in yield development within the branch becomes visible.

Among the winners in this crowding-out competition are the Japanese groups Toyota, Honda, Nissan, and the Europeans Renault and BMW. If one neglects the two statistical strays, Nissan and BMW with their high operative losses in 1999, one discovers a clearly rising trend up to 2003. This positive upwards trend seems however not to have continued in 2004, according to the company reports available so far.

The groups GM, Ford, PSA, Fiat, DaimlerChrysler and Volkswagen, however, are among the losers in the global crowding-out competition. Following a relatively high-yield phase at the end of the 90's, the profit situation has deteriorated rapidly in the last few years and has stayed at a stagnant average level of approx. 0.5% (Fig. 14) since 2001. According to company data so far available nothing changed in 2004. On the contrary.

Three things are remarkable about this differentiation in profits:

1. For one thing, it is not only the premium manufacturers who are in the "winners" group - as may have been expected – but producers with sales focuses in the middle and lower market segments. It is obviously management skills and not market sectors, which are decisive for the long-term success of a company.

2. On the other hand the firms in the "winners" group have in common that the manufacturers have a comparatively slim portfolio of brands. The opposite is true of the "losers" group.

3. Up until 2003, at least, the profit trend split between winners and losers widens tendentially: the "good" get better and better and the "bad" get worse and worse. The winners have obviously exploited the chances offered by globalization better.

Fig. 14. Trend developments of net sales yields, in per cent

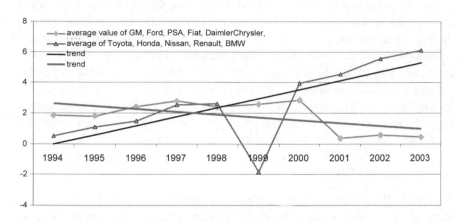

source: annual reports, ComStock, IWK calculations

As already explained, the outlined development is valid up to 2003. Since then the global crowding-out competition has caught up with the "winners" group, too, according to current information.

1.8   Passing the loss on to the components suppliers

The pressure on costs and yield is passed on from the automobile manufacturers to the upstream stages of automobile value creation.

The supply industry is in the main focus here. The manufacturers attempt first of all to indemnify themselves against their shrinking profit margins. The price pressure on component suppliers thus increases considerably. The toughest price negotiations and the most rigid cost objectives come from the manufacturers whose profit situation has worsened most, as experience shows. This automatically and immediately mobilizes those manufacturers who have better margins, but cannot afford to miss the opportunity offered by the ailing competition.

The suppliers are expected not only to develop new technologies efficiently and offer them exclusively, but also to finance themselves in advance. In addition they are expected to continually optimize the manufacturing processes of existing products, in order to generate regular price reductions. In this manner, the supply industry serves as a cost buffer for the manufacturers, who sometimes exploit their market power as buyers to such an extent that suppliers are forced to accept orders with no cost coverage, in order not to lose their customers. The OEMs are able to exploit their position with regard to the suppliers so mercilessly because the concentration process is much more advanced at the manufacturer level than at the supply level. Because of the small number of remaining OEMs and the increased use of identical parts for different models, the number of orders given to suppliers is sinking, although order volumes are increasing. This inevitably leads to more intense competition among the suppliers, spreading through the entire components supply branch from top to bottom. It leads, in addition, to higher capacity risks – as a result of the higher order volumes – if the OEMs have to fight sales problems and cannot keep up with their call-off plans.

The result of this is that the concentration spiral twists permanently at supplier level, the weakest leave the market and are taken over by larger organizations, because market power can only be built up against the demand power of the few OEMs if the supply structure is oligopolistic.

The persistent pricing pressure (and the resulting fight against collapsing margins) is only one of several financial challenges facing the components suppliers at the moment. Additionally, manufacturers are also increasingly reducing their own vertical range of manufacture and passing more and more development risks on to the components suppliers. This means that the quantity and complexity of development projects increases for the suppliers – and thus also their capital requirements.

The component suppliers thus face a dilemma: on the one hand they must maintain their leading role in costs and technology in order to stay competitive. On the other hand, financing the necessary investments presents a growing challenge in view of the fact that the banks have increased their credit security requirements considerably. The manufacturers' tense sales situation and the resulting pricing pressure on suppliers thus reinforce the financial bottleneck faced by these suppliers.

The manufacturers' price expectations can often no longer be met if they produce in Germany. The only solution is to relocate production into low-cost foreign countries. Thus German component suppliers are often forced to exploit the cost advantages offered by producing abroad, despite the connected risks and disadvantages.

A large number of suppliers has already shifted parts of their production to cheaper countries, regardless of the OEMs. For many, this was the only way to remain "listed" with the OEMs and at the same time protect their margins.

1.9   Conclusion: branch with increasing profit pressure

In the global automotive industry all the enterprises involved are confronted with far-reaching sales and income problems. The concepts with which manufacturers previously attempted to overcome the drop in sales, with hardly manageable variant and model campaigns, have so far not eased the situation. This is however not particularly surprising, as there is a simple reason behind it. As all the OEMs are doing the same, the one which shrank from this policy would be at a disadvantage. So they all join in – and block each other. Thus it is not so much aggregate demand for vehicles which is stimulated, but mainly cost and margin pressure on all of them, even if it is not equally painfully for all of them.

Imaginative discount battles in the big volume markets also remain without positive results. They increase marketing costs and put pressure on profit margins, but they do not lead to a recovery of the market. Such discount campaigns thus make a big contribution to the economy as a whole, because the manufacturers give a part of their "producer's surplus" to the consumer, but they do not result in any lasting commercial success.

The branch is caught in a dilemma between *structural market saturation* and *merciless crowding-out competition*, which will inevitably lead to the withdrawal of the weakest competitors. The only feasible solution for the manufacturers is the consistent reduction of costs and overcapacities, in order to stay profitable despite the lack of growth.

Those are the aggravated conditions which apply to all the manufacturers. But to blame only the markets and the competition for shrinking net profits would be too simple. As the positive developments made by a few manufacturers against the trend show, those others with worse performance must have additional home-made management and sales problems. Profitable producers are obviously better managed, while other companies make strategic mistakes, for example in multinational concern policy, models policy or quality assurance. According to knowledgeable observers of the branch, VW's new Golf was too expensive and the entry into the luxury class with the Phaeton was planned badly. GM's strategic mistakes in Opel model policy are sufficiently well known. DaimlerChrysler had to withdraw from its commitments to Mitsubishi after suffering high losses, and BMW had gone through the same drill a few years earlier. And there are plenty more such examples.

It is also unmistakable that just as profit pressure has increased in the automotive industry, so too have quality problems. The reason was usually that measures to reduce purchasing costs and rationalize processes were too rigid. Cost-intensive callbacks were the result for (almost) all manufacturers.

It is difficult to come to a reasonable judgement of the branch. There are the problem cases described above, but also the manufacturers such as Porsche, Audi and BMW, who are extremely successful despite the adverse conditions in the market. The combination of attractive products and high image factor, the concentration on core business and expanding core brands, and efficient cost management would appear to guarantee success even in difficult times.

The fact that mass producers can also work profitably in saturated markets is proved by the highly profitable Japanese producer, Toyota, now focussing very successfully on the highly sensitive European market. The same applies to Renault and – within limits – to PSA with their innovative design trends and diesel technologies.

The result of the actual situation analysis can be summed up as follows: The entire automobile branch throughout its whole added value chain is going through a lasting upheaval. The essential task for each of the remaining OEMs is to improve their own competitiveness so they can continue to act independently in the global crowding-out competition. Wherever the exhaustion of a company's own cost reduction potential is not sufficient, manufacturers are now forming strategic coalitions, which would have been unthinkable only a few years ago.

It remains to be seen whether this is enough for a consolidation of the crowding-out competition among the remaining manufacturers. *For why should the strong producers go easy on the weak ones, if, as a result of the weak ones withdrawing from the market, they are able to generate the profitable growth they need?*

It is against this backdrop that the next chapters will attempt to describe the challenges facing the German automotive industry, the future general conditions in the automobile markets of the world, and the probable reaction patterns of today's big market participants and their suppliers.

2 The Western European automotive industry: cost stress and profit pressure

2.1 External pressure factors

2.1.1 Cost gap in international comparison

The international competitiveness of an industry is determined on the one hand by "hard" price and cost factors, and on the other hand by "soft" non-price factors such as product quality and reliability, innovation capacity, customer service, reliable delivery, system security, management skills, environmental acceptability, etc.

This is valid to a high degree for the Western European, especially German automotive industry. As its continuously strong export position demonstrates, it has in the past always been successful in compensating for the competition disadvantages of high costs incurred by engineering- technological products and processes. The excellent image and proverbial quality of the German motor industry in the "soft" factors were the key here.

In the meantime the signs have been increasing since the year 2000 that this superior position is slowly but surely crumbling – partly because of the branch's own failings and mistakes, but to a greater extent because of the changed conditions in global competition. On the one hand, foreign competitors are catching up fast in the "soft" factors that means the competitors are getting better, while the numerous call-backs in recent years signalize that Western European producers are obviously unable to properly defend their prime position.

It is however, more important that virtually over night in the course of globalization competitive production has become possible at locations which until recently were inaccessible for political reasons. Some of them may have been of interest to the tourist trade, but not to industry. A competitive gulf has suddenly opened up, which cannot be blamed on companies themselves, but on the fact that production locations may suddenly be (and want to be) bound into a global division of labour, which – for whatever reasons – used to be well beyond the planning horizon.

Looking at the pressure factors in individual national economies, the labour costs, as cost factor no. 1, are the focal point. Apart from the framework presented by the law, by fiscal and economic policy, and by society, Western Europe is suffering mainly from labour costs which are high by international comparison.

Therefore the individual cost factors are first analysed below, to clarify the competitive positions, or rather the deterioration of the competition in the automotive industry. Following that an assessment of Germany as a leading Western European industrial location, with reference to the international competitiveness of the automotive industry, can be made.

2.1.1.1 Labour cost per hour

Labour costs represent a substantial factor in the calculations of companies, and are defined here as gross income of employees (incl. social security costs). As shown in the results of numerous surveys, the level of these costs, particularly in the components supply industry, is decisive for the choice of production locations (for about ¾ of the companies surveyed).

Fig. 15. Cost structure of entire added value in the automotive industry

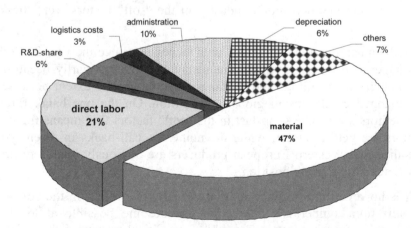

source: Automobile Engineering Partners, IWK presentation

Examinations of the cost structure of the branch (Fig. 15) show that labour costs represent a proportion of the overall value added in automotive industry of hardly one fifth. That means it is only half as big as the proportion of material costs (47%). *Does that mean that labour costs are negligible?*

Such a conclusion – although often used in public discussion by employee organizations, union officials and works councils to trivialize the high-wage problems in the West European economies – is wrong. The opposite is true! The fact is that the cumulative wage costs in the whole value creation chain of the product automobile must be included in the calculation. Because wage costs are also included to a great extent in material costs and other external services which are used in the manufacturing process (IT, development services, advisory costs, etc.). In fact more, the nearer one gets in the value chain to the tier 1 suppliers. Here the share of cumulative wage costs in value added amounts to 50% and more.

In the whole automotive value chain the share of wage costs in the final product is at least 65%, according to expert estimates. If one considers the wage costs incurred by suppliers and adds to those the R&D, logistics and administrative costs, which are also to a great extent made up of wage costs, it becomes apparent that wage costs make up over 75% of overall manufacturing costs in the automotive industry.

Both statutory and voluntary additional personnel expenses (employers' contributions to social security, continued wage payments in case of sickness, vacation pay, individual capital formation schemes, shift differentials, "desert premiums", etc.) are included, which almost doubles the actual hourly wage costs. But in 2004 a turnaround was initiated along a wide front at both managerial and statutory levels. The motto is the reduction of the high non-wage labour costs in order to break down the barriers to new recruitment. Those involved all agree about the direction, just not (yet) about the extent of these cuts.

External inflation shocks in raw materials such as aluminium or steel only appear to have a positive effect on these relations, if OEMs are unable to compensate higher raw material costs by reducing wage costs, but have to pass them on to the customer in the form of higher sales prices or – as is currently the case – have to absorbing them through shrinking profit margins.

Considering the significance of wages as a cost factor in national-economic value calculations, the absolute size of these costs plays a central role in determining competitiveness in the international comparison.

Fig. 16. Labour costs in the West German manufacturing sector, in Euro
per employee and year

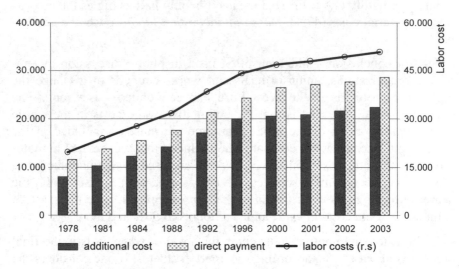

source: IW Cologne, IWK presentation

In the West German manufacturing sector in 2003 wage costs raised
above the benchmark of 50,000 Euros for the first time, at 50,930 Euros
per employee, reckoned pro employee in full-time units. Additional personnel costs also reached a new peak level, at 22,350 Euros. In the period
form 1978 to 2003 they rose more rapidly than direct payments, at an average yearly rate of 4.1%. The proportion share of additional personnel
costs rose by 8 percentage points in this period and in the end reached
78.2% (Fig. 16)

The competition problem of labour costs becomes particularly evident
in international comparison. The manufacturing sector (Fig. 16), but above
all the German automotive industry (see Table 3 and Fig. 17) have to cope
with the highest labour/wage costs of the twelve most significant industrialized countries.

Table 3. Wage costs in automotive industry in international comparison
(Euro/hr.)

country	1990	1998[*)]	1999	2000	2001	2002	2003	2004
Germany **)	20.22	30.89	31.60	32.22	32.28	32.13	32.75	34.00
France	12.81	18.53	19.65	20.51	20.98	21.51	21.45	22.69
Italy	13.95	15.96	15.39	15.93	16.25	16.36	16.81	17.36
The Nether-lands	13.83	19.68	20.26	21.81	22.91	22.98	23.71	25.00
Belgium	16.50	23.16	23.84	24.28	25.40	26.30	26.66	27.06
UK	12.21	19.05	20.01	23.03	23.03	24.01	22.49	23.60
Sweden	20.85	20.40	20.95	22.77	21.03	21.96	23.44	24.13
Spain	13.16	14.85	15.36	16.09	16.68	17.24	17.80	18.24
Portugal		7.50	7.86	8.23	8.54	8.78	9.02	9.24
Austria	14.94	22.34	22.76	22.61	23.59	23.96	24.48	25.09
U.S.A.	16.09	22.75	24.88	29.89	31.72	31.90	27.65	25.49
Japan	13.21	21.62	28.63	35.87	34.04	31.58	28.81	27.28

*) until 1998 incl. based upon the annual average of official ECU-exchange rate.

**) until incl. 1997 old federal states, level: January 2005.

source: VDA

Fig. 17. Wage costs in the automotive industry in international comparison
(Euro/hr.)

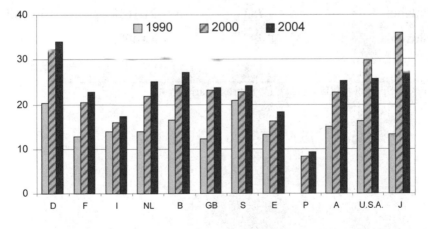

source: FERI

Wage costs per hour in the German automotive industry, at 34 Euros (without social expenses) are about 25% higher than in Japan, at 27 Euros, and 267% higher than in Portugal (9 Euros). Taking as a basis the more broadly-based comparison of labour costs with the new EU countries in Eastern Europe or other membership applicants, the unfavourable comparison turns out even worse for the location Germany (compare Table 5, Table 4 and Table 6).

Table 4. Labour costs in the new acceding countries, 2003

country	labour costs *(in Euro per month in industrial and service providing sector)*	labour cost index *(in % of German level)*	amount of employees
Latvia	357	9.60%	1.1 mil.
Lithuania	487	13.10%	1.6 mil.
Slovakia	565	15.20%	2.6 mil.
Estonia	608	16.40%	0.7 mil.
Hungary	764	20.60%	4.2 mil.
Czech Rep.	777	20.90%	5.1 mil.
Poland	783	21.10%	16.9 mil.
Slovenia	1.497	40.40%	1.0 mil.
Germany	3.710	100.00%	39.8 mil.

source: Spiegel (2005-02-21), IWK presentation

Table 5. Labour cost in the manufacturing sector in 2002

country	employees´ labour costs	Among them		workers´ labour costs	additional cost quota *in % of direct payment*
		direct pay-ment	*Addi-tional staff costs*		
		per hour in Euro			
Slovenia	9.01	5.38	3.63	-	67.5
Czech Rep.	5.03	2.75	2.28	4.17	82.8
Hungary	5.03	2.82	2.21	3.97	78.5
Poland	4.49	2.82	1.67	3.61	59.4
Slovakia	3.46	2.02	1.44	2.91	71.3
Estonia	3.19	2.09	1.11	-	53.0
Lithuania	2.83	1.86	0.96	-	51.7
Latvia	2.29	1.59	0.69	-	43.5
Romania	1.46	0.86	0.6	-	70.0
Bulgaria	1.23	0.73	0.5	-	68.8
West Germany	31.67	17.84	13.83	26.36	77.5
East Germany	19.09	11.65	7.44	16.43	63.9

source: Institute of German Economy Cologne on basis of national data.

Table 6. Labour cost in road vehicle construction (per man-hour)

	2000		2002
	in EUR	*manufacturing sector =100*	*in EUR*
Lithuania	2.01	92.2	2.11
Latvia	2.34	91.1	2.58
Poland	4.56	115.4	5.18
Slovakia	2.72	89.2	3.09
Czech Republic	4.37	122.1	6.14
Hungary	4.64	126.1	6.34
West Germany	39.15	130.0	41.16
East Germany	20.47	114.6	21.88

source: Institute of German economy Cologne on basis of national data.

As presented in the following Fig. 18, one manpower hour costs industrial enterprises in Germany a good 26 Euros. This means German employees are among the most expensive in the world. Only Norway and Denmark show higher labour costs, but they have no automotive industry, as it is well known. Important countries in competition with Germany, such as the USA, Japan, France, Italy, and the Netherlands enjoy a unit cost advantage of as much as 30%.

Fig. 18. Wage cost in industrial production in international comparison, (Euro/h.)

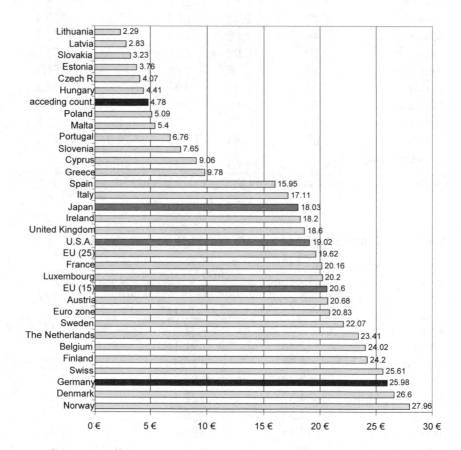

source: Gesamtmetall

As soon as the average direct payments in the countries being considered are compared, a distinct east-west differential becomes apparent. Only Portugal distinguishes itself by a very low wage level, even though wages there are still more than twice as high as in Slovakia, for example.

Comparing the proportion share of additional costs, the differences between east and west Europe are not so obvious. A few west European countries, such as Ireland and Denmark, are even only on a level with Slovakia, whilst Turkey takes a prime position in front of Italy, for example (Compare Fig. 19).

Fig. 19. Additional staff costs in international comparison, in % of direct payment

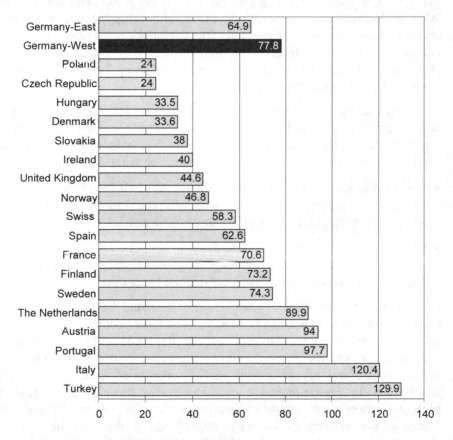

source: Gesamtmetall (2004), IWK calculations

But all this is only valid for the relation between direct and indirect wage costs. Considered in the absolute, additional personnel costs mirror in principle the regional direct payment profile. With the exception of Portugal, additional costs in West Europe are at least four times as high as in the east European Countries. However within the west European countries changes in the ranking as related to direct payment can be observed. Here Great Britain is next to Portugal at the bottom of the league, whilst the West German additional costs are right at the top.

West German additional personnel costs in 2003 were at an average of 22,350 Euros per full-time employee. Despite a further rise in eco-tax, the introduction of which was supposed to relieve pension insurance, this represented a new peak level. In contrast, the newly-formed German states enjoy distinct advantages. The absolute level of additional personnel costs there is a good 20 per cent below the West German level. Thus East Germany is in a more competitive position in comparison with the rest of Western Europe, but in relation to the neighbouring Eastern European countries however, still worlds apart. This explains among other things the economic policy difficulties in building up the East German economy: state seed subsidies cannot substitute the long-lasting cost disadvantages compared to the neighbouring countries.

The low labour cost level will be a positive location factor in the low-wage areas of central and Eastern Europe and Asia for many years, judging by the emerging global economic trends. Yet before any decision to shift location to these regions, two things must be taken into account:

- On the one hand, production levels there are often lower than in West Germany, so that the unit labour costs differential turns out lower than suggested by the differences in labour costs (see at 2.1.1.2.).

- On the other hand, an appropriate development in exchange rates could partially exhaust the competitive advantage of low labour costs, because the low-wage countries are (still) outside the Euro zone.

Moreover, for the sake of completeness it should be pointed out that other very important location factors, such as legal security, political stability, infrastructure, etc. also play an important role in location calculations, and not only the differences in labour costs. When faced with additional logistic and organizational costs a change of location is hardly worthwhile for capital-intensive manufactures or those which require a high proportion of specialists and skilled workers, which are usually also scarce or non-existent in low-wage countries, and therefore have to be "imported" from Germany. This applies less to the scientific-technical field, in which many people were educated under socialism in the central and Eastern European states, for lack of alternatives in the humanities. This is illustrated by the increasing relocation of development departments from German suppliers to Eastern Europe, for example Romania, or even to China.

This may be an additional incentive, little noticed by the public eye, to outsourcing and relocation of development departments to low-wage countries. Because it is just the lack of engineers which is a key problem for Germany. German industry suffers from an acute lack of skilled workers, as the number of engineers has dropped sharply in the last few years. There is an estimated yearly lack of 20,000 engineering graduates. According to the Association of German Engineers (VDI), the demand for engineers, especially in the fields of mechanical engineering and electronics, is increasing by 6% per year. [15] From the current viewpoint this demand cannot be met in Germany in the foreseeable future.

2.1.1.2 Productivity per man and hour

From an economic point of view a *high wage level per se* is not a problem in a national economy, as it depends on the performance achieved at this wage level. If there is more gain to be made at one location than elsewhere – that is, if productivity is better – then performance can be honoured at a higher level. For decades this was the secret to success for the German economy's great international competitiveness, despite high wage costs.

But is this argument still true of Germany as an industrial location? Unfortunately not! A more detailed analysis shows that in the meantime in many regions of the world enterprises are able to operate certain kinds of standardized production with high productivity levels and simultaneously low wage levels. For example when they invest in low-wage countries and take their technology and production know-how with them. Thus when decisions have to be made about a new location, the differences in wage cost are often the deciding factor.

---

[15] Conf. HAWK 2015 (2003), p. 15

Fig. 20. Unit labour costs internationally in the year 2003, Germany = 100

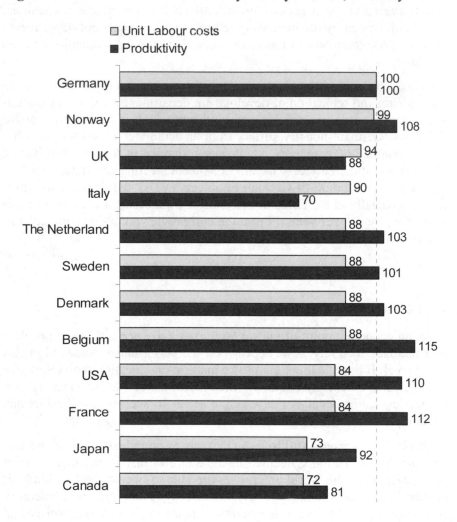

Manufacturing sector, wage unit costs: Relation of labour cost per employee hour in price and at 2003 exchange rates to productivity;

Productivity: gross value added to manufacturing cost per person in dependent employment hour in prices and at 2003 exchange rates, in the U.S.A. and Japan gross added value for market prices. Original data: German Federal Bank, OECD, U.S. Department of Labour

source: iwd (2004d)

Moreover the oft-cited German productive edge is also hardly more than self-satisfaction and wishful thinking (compare Fig. 20). In seven out of eleven important industrial nations productivity was higher than in Germany in 2003. This may have gradually improved in 2004 as a result of the lasting rationalization efforts in German industry, but definitely not in the automotive industry. The number of employees in the branch even increased by about 4,500 according to the VDA, despite stagnating production figures. Thus productivity can be expected to have fallen rather than risen as a result of the lack of volume expansion in the branch. Individual exceptions at certain OEMs, such as Porsche, or suppliers, such as Bosch or Continental confirm the rule. Basically it is true that progress in productivity which does not go hand in hand with increases in production can only be realized through internal rationalization measures and job reductions, and is therefore difficult to implement.

Below the line, productivity in Germany in 2003 was only 1% above the average of the other eleven countries. Considering this mediocre performance it is clear that Germany's unit labour costs – that is, the relation of labour costs to productivity - aren't exactly brilliant. On average, the competing industrial countries enjoyed a labour cost advantage of 16 to 20% over German industry. The gap is particularly large between Germany on the one hand and Japan and Canada on the other, where labour costs per unit of production were a good 27% lower. Even France and the USA were able to beat Germany by 16%.

Hence it follows that no other industrial nation burdens its production with labour costs so heavily as Germany. The difference to competitors from highly developed industrial countries, such as Japan, the USA, France, and Great Britain amounts to 20-25% grosso modo. Compared with the new EU membership countries in the East, the disadvantage is significantly greater.

2.1.1.3 Labour time/ labour volume per year

Germany not only holds a prime position in wages, but also in the reduction of working hours, which many European industrial countries have seen in the last few decades. Considering the intensification of the global competition for jobs, the current discussion of a return to longer working hours is long overdue.

It is above all the motor industry which today is faced with an unfavourable mixture of relatively low standard working hours and high labour costs. This double burden exists in the same manner in no other country. The situation has become even more volatile through the entry of the central and Eastern European countries to the EU (see at 2.1.2.1.).

The 35-hour week has been valid since 1998 for the whole of West Germany in all collective agreements in the metal and electrical engineering industries. The steel industry had in 1994 already introduced the shorter working week by collective agreement. Other branches followed the metal industry only a short way. Even so, the average agreed standard working week of all branches in Germany is now only 37,5 hours (39,1 in East Germany). By orientating itself to a 35-hour week Germany has taken a by-path, along which only a few other countries have followed. Among them are the Netherlands, Denmark and in particular, France. Since 2002 the 35-hour week has been the statutory standard in France, after the introduction of the shorter working week was promoted with financial incentives for companies in 1998. However, in most other EU partner countries the length of the working week has remained mostly unchanged at least in the last 10 years.

Table 7. Standard minimum and maximum holiday in the West European metal and electrical engineering industry, in days

United Kingdom	23	25	25	25	25	25	25	25	25	20-25
Italy1)	25	25	24	24	24	24	24	24	24	20-25
The Netherlands	22-31	22-31	23-32	25-38	25-38	25-38	25-54	25-54	25-54	25-54
Norway	20-25	21-26	21-26	21-26	21-26	21-26	23-28	25-30	25-30	25-30
Austria	20-25	24-29	25-30	25-30	25-30	25-30	25-30	25-30	25-30	25-30
Portugal 1)					22	22	15-22	15-22	15-22	15-22
Sweden	25	25	25	25	25	25	25	25	25	25
Swiss	18-25	20-25	21-26	22-30	25-30	25-30	25-30	25-30	25-30	25-30
Spain 2)		26	26	24	24	26	26	26	26	26
Turkey	12-24	18-26	18-26	18-26	18-26	18-26	18-26	18-26	18-26	18-26

1) for employees with minimum 1 year staff membership in the same firm
2) Saturdays run as working days and are counted.   *) provisionally
sources: CEEMET - Council of European Employers of the Metal, Engineering and Technology-based Industries,   statistical exchange

source: Gesamtmetall (2004)

The problem of Germany's unfavourable weekly working hours accentuated further by a generous vacation entitlement (compare Table 7). At 30 days this is well above the EU average of just under 24 days. Even if the number of public holidays in Germany, at 10.5 days in 2003, which don't regularly fall on a weekend *(when leave is taken to make up a long weekend –bridging days)* is not is not any higher than average (EU: approx. 11

days), the combination of vacation entitlement and bridging days results in a disproportionately high level of absence from the workplace for German employees.

Table 8. Theoretical standard annual working hours[1] in the metal and electrical engineering industry, in hours

Country	1960	1985	1990	1995	1998	1999	2000	2001	2002	2003	2004[2]
Germany West	*1808*	*1698*	*1623*	*1573*	*1556*	*1563*	*1540*	*1535*	*1535*	*1542*	*1570*
Germany East	-	-	-	1724	1689	1697	1672	1667	1667	1674	1705
Belgium	1775	1721	1709	1702	1709	1709	1702	1709	1709	1709	1717
Denmark	1836	1804	1681	1672	1662	1676	1650	1628	1628	1621	1658
Finland	1872	1816	1716	1716	1724	1732	1756-1716	1756-1716	1748-1708	1748-1708	1772-1732
France	1872	1740-1717	1748-1725	1740-1717	1748-1725	1763-1740	1582-1561	1589-1568	1582-1561	1575-1554	1610-1589
UK[3]	1856	1778	1778	1771-1680	1778-1687	1771-1680	1771-1680	1771-1658	1880-1680	1880-1687	1888-1695
Italy	1832	1784	1768	1736	1768	1776	1736	1768	1768	1752	1776-1736
The Netherlands	1856	1752-1680	1744-1672	1720-1616	1736-1632	1744-1640	1728-1635	1832-1600	1728-1496	1728-1496	1752-1520
Norway	1864	1840-1800	1725-1687	1725-1687	1740-1703	1747-1710	1725-1688	1710-1673	1695-1658	1703-1665	1725-1688
Austria	1856	1802-1762	1727-1689	1717-1679	1736-1689	1744-1706	1716-1677	1731-1693	1739-1700	1731-1692	1754-1716
Portugal	-	-	-	-	-	1816	1816	1896-1840	1896-1840	1896-1840	1920-1864
Sweden	1816	1800	1800	1803	1772	1784	1766	1750	1746	1729	1758
Switzerland	2012	1949-1907	1856-1816	1840-1776	1848-1784	1848-1808	1816-1776	1824-1784	1808-1768	1808-1768	1848-1808
Spain	-	1826	1800	1784	1792	1808	1768	1776	1778	1776	1784
Turkey	2217	2115-2055	2115-2005	2113-2059	2111-2051	2010-2029	20S1-2021	2130-2070	2126-2066	2111-2051	2134-2074

1) annual labour time determined by tariff and legal settlements for full-time employees; 2) provisionally; 3) revised since 2002; source: Gesamtmetall

As a consequence, Germany is right at the bottom of the European scale of collectively agreed annual working hours in the metal and electrical industries, although working hours have been rising slightly since 2001 as a result of the economic crisis (Table 8). In 2004 standard yearly working time in Germany amounted to about 1640 hours.

It is worth pointing out in this context that the USA succeeded in stopping the trend toward effective reductions in working hours as early as the beginning of the 80's, whilst Germany's disadvantage in this area has continued to grow until today - despite a collectively agreed raising of working hours (Fig. 21)

Fig. 21. Development of effective annual labour time in Germany and the U.S.A.

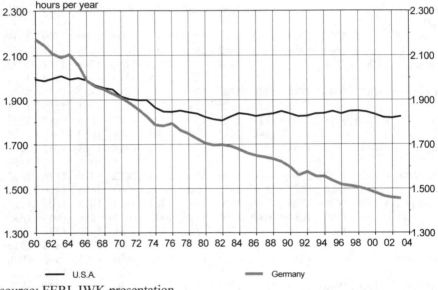

source: FERI, IWK presentation

2.1.2 Intensification of global location competition

2.1.2.1 EU- enlargement in the East

As a result of the entry into the EU of the five central and Eastern European countries Poland, the Czech Republic, Slovakia, Hungary, and Slovenia, as well as the Baltic States, completed on May 1st, 2004, global location competition has intensified further. It is true that the previous development of foreign trade and direct investment (Fig. 22) shows that many western companies have been active in the central and Eastern European countries since the 90's, and have been utilizing the existing potentials, but their EU entry signifies a renewed improvement in general conditions for the new member countries.

Fig. 22. German direct investments in East Europe

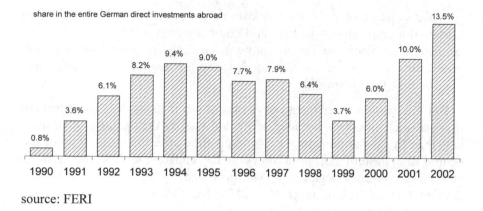

source: FERI

Growing legal security, the assumption of EU standards and increasing political and economic stability will particularly move those small to middle sized enterprises which until now have been cautious, to invest more in the central and Eastern European countries, in order to utilize the considerably lower cost levels, build up new production capacities and increase local sales. This will lead to a middle term deepening of commitment to new member countries on the part of Western Europe and Germany in particular. At the same time, the economic linking of these countries to Western Europe and above all Germany will be further intensified. For many German enterprises the entry of these countries into the EU will be an impulse for expansion and an increase in activities in the new member's markets, first of all to produce for export to Western Europe at lower costs, but with increasing prosperity in these countries also for the local market.

It was the automotive industry especially, which rapidly assumed a pioneering role in the development and opening up of the markets in Eastern Europe after the collapse of the Eastern Block.[16] In particular it was the German manufacturers who, due to their geographic and cultural prox-

---

[16] Conf. IWK (2004a).

imity, functioned as pioneers in this sense. As a result of the eastward enlargement of the EU the automobile market in Eastern Europe will play an increasingly important role for German OEMs and suppliers. For some more as a sales market at first, but for others as a low-cost alternative production location to Germany. The share of central and Eastern European states in German automobile exports is at present 6.5%, in 1995 it was only 3.6%.

A direct shift of production locations of German OEMs to Eastern Europe through shut-downs here and complete reconstruction there has so far not taken place. So far, investments in capacity growth for existing production lines or the addition of capacity for new models and variants have been in the foreground.

But this type of investment obeys other laws than those in other growth regions, such as China, because it is ruled less by the opening up of markets than by cost arguments. Only a short time after the opening up of the central and Eastern European states at the end of the 80's, the German automotive industry began to invest there heavily. As soon as 2004 production was 1.4 million units, by 2007 capacities will almost double, according to current plans (Fig. 24).

Both the Czech Republic and Slovakia are today already among the countries with the highest per capita automobile production. this build-up of manufacturing capacity is problematic from the German point of view, above all because the suppliers join the OEMs in moving east, so far as they are not there already. Continental Plc. is an example of this, having already shifted most of its tire production to Eastern Europe. Their development department is following now. [17]

These investments are mainly made to supply the nearby west European market (and oversees), and at the same time to be able to produce more cheaply than in Western Europe. In 2003 30.5% of the passenger cars imported into the EU-15-countries already came from Eastern Europe. [18] New investments and capacity expansion, already announced, on the part of Asian manufacturers will inevitably lead to greater production capacities in Europe. It would seem that the sale of this extra volume in Eastern Europe and Russia alone can be ruled out in the foreseeable future – the main goal of this capacity expansion is first of all to supply the large market of West-

---

[17] For a current overview of the number of manufacturing plants of German automobile enterprises in middle and East Europe see VDA (2004b), p. 69-88.

[18] Source: Eurostat

ern Europe. But at the same time that means that the destructive competition on the saturated Western European market will become fiercer.

The significance of Eastern Europe in the short and medium term will increase above all as a production location and less as a sales market. The average income, which is still distinctly lower than in Western Europe, and the lower buying power in the Eastern European countries mean that the relatively highly-prices of western manufacturers' vehicles are (still) exorbitant for many people. With the increase in prosperity in the population which can be expected as a result of the EU enlargement however, Eastern Europe will gain greater importance as a sales market for new vehicles. With relation to the passenger car density there is however a not insignificant backlog of demand, compared with western levels.

Fig. 23. Passenger car stock per 1.000 inhabitants (2001)

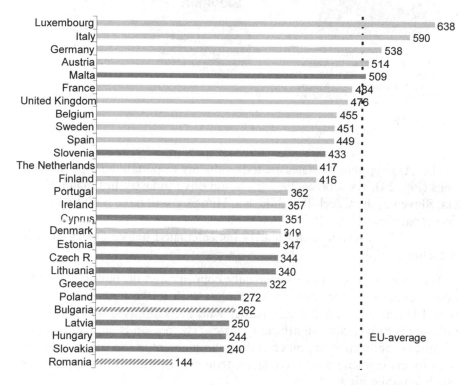

source: Federal Statistic Office

Fig. 24. Production locations in East Europe 2004

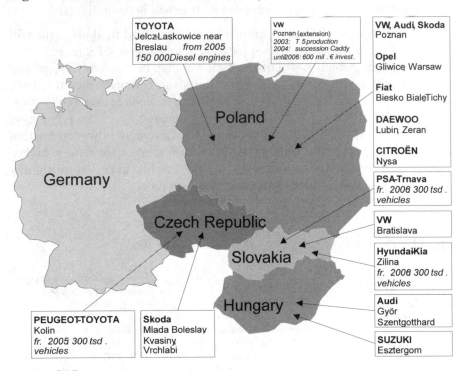

source: CAR

The EU enlargement states are today already important automobile locations (Fig. 24). As soon as 2007 the automotive industry in Poland, Slovakia, Slovenia, the Czech Republic, and Hungary will have built up production capacities to 2.6 – 2.7 million vehicles, corresponding to about 50% of German vehicle production capacity (as calculated before Opel's capacity reductions).

The *manufacturers* have so far only concentrated on building up additional capacities rather than on a complete shift to the new Eastern European EU states. It is a different matter with the *suppliers*, whose capacity build-up is progressing significantly faster. The reason is simple: The possibilities of relocation are much easier to implement in the supply industry, with its manageable workforce sizes, than for the OEMs, which are very much in public view.

When choosing a manufacturing location suppliers pay particular attention to low wage and production costs and good availability of qualified and highly motivated employees. Here the central and eastern EU states offer clear advantages over the traditional locations in Germany and Western Europe. In the medium term it can be expected that Western European suppliers who do not manufacture in Eastern Europe will be faced with considerable competitive disadvantages due to their cost structure, and will have to leave the market sooner or later.

Fig. 25. Labour cost index of the EU-entry countries relative to EU-15, 2004

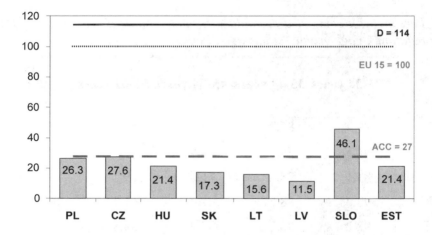

source: FERI, IWK presentation

Labour costs in the Associate Candidate Countries (ACC) of the EU are only 20% of the level in the old EU, while Germany is at a level of 115%. Only in Slovenia, where many western companies, in particular from the automobile branch, are already active, are labour costs meanwhile twice as high as the average in central and Eastern European states.

But in addition to labour costs, the flexibility, qualifications and attitudes of employees are important arguments for western investors when assessing a location. Actually the question is not only about whether the machines can operate round the clock – which is possible in Germany, Eastern Europe or China. It is also about whether a qualified personnel is available round the clock, who can operate and monitor these machines, i.e. it is about the training of these workers.

The population of working age in the new EU countries has a distinctly higher educational level than in Western Europe, and that in all age groups. In Table 9 the proportion of persons of a particular age group is shown, who have finished schooling successfully at least to secondary II level. In Germany this corresponds to a secondary school final. In the new EU countries this proportion is 78%, in the EU 15 however, only 64%. The differences between the individual central and Eastern European countries here are not great, all the countries are clearly above the EU 15 average.

Table 9. Educational achievement by age groups

Persons with completed secondary II education as a proportion of the population, in age groups, status quo 2000

	25-34 years	35-44 years	45-54 years	55-64 years	Total (25-64 years)
PL	89.4	87.2	77.1	56.2	79.7
CZ	92.5	89.4	83.9	75.7	86.1
HU	81.1	78.5	71.3	39.2	69.2
SK	93.7	88.6	81.4	61.1	83.6
LT	92.2	96.2	87.4	54.8	84.9
LV	89.7	91.3	83.4	66.2	83.5
SLO	85.5	78.2	70.8	60.9	74.8
EST	90.8	91.9	85.8	66.3	84.7
Ø ACC	87.7	85.8	76.8	56.4	78.3
EU 15	74.2	68.1	58.7	48.0	63.5

source: Eurostat

When enquiring about the sufficient availability of employees, attention must be paid to the fact that the central and Eastern European states, with a total of 33.3 million employed, have only 6 million fewer workers than Germany (39.8 million). But also that, with the exception of Poland, each country on its own has only a relatively small population, and therefore only a relatively small number of workers (Fig. 26). Thus there is a natural limit to the expansion of production sites in these countries, based on the labour potential, because with increasingly short numbers of workers in free competition the wages and thus costs for companies will rise. That at least partly explains the relatively high wage level in Slovenia, which has fewer than one million workers and a low unemployment rate, and thus a certain shortage of qualified workers.

Fig. 26. Working population in the ACC-states, 2004

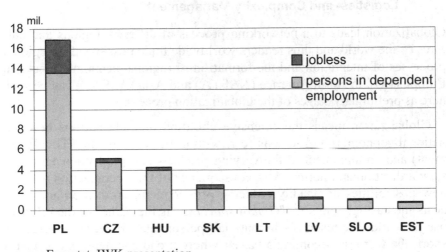

source: Eurostat, IWK presentation

A further significant location factor for suppliers, above all systems and module suppliers, is proximity to the OEM. Because these suppliers are meanwhile accommodated increasingly often on the factory premises of their OEM clients for cost reasons, sometimes even in the same production hall, the potential for relocation is directly linked to the location of the end producer and is roughly zero. New options for the supplier only appear if the OEM itself relocates its production.

This is however not valid for direct component suppliers or the suppliers' suppliers. Depending on the kind of business relationship, spatial proximity can be substituted by modern transport and logistic systems. Where there is a good infrastructure or a favourable geographic situation, suppliers can transport their products without problems to their customers, end producers or other suppliers, even over great distances. A just-in-time or just-in-sequence supply is hardly possible this way of course, but in many cases – depending on the product – this is not always necessary.

Other location factors, such as low taxes, state investment incentives or a stable political framework are, according to surveys by Ernst&Young[19], significant for firms in the supply industry, but not of decisive importance. The direct costs are definitely the dominating factor.

2.1.2.2 Increasing Globalization as a result of improved Production-, Logistics- and Complexity Management

Globalization leads to a networking process of different regions and cultures of the world and thus renders worldwide trade possible. The regional processes of integration and the formation of regional economic blocks in Europe (EU), in North America (NAFTA) and Asia (ASEAN) can be seen here as preliminary stages of the globalization process.

Globalization means the extensive abolition of trade barriers between national economies and regions (e.g. within the scope of the EU enlargement) and an alignment of the varying products and production forms by standardization and norms. As a consequence of the liberalization of trade new possibilities of steadily reducing procurement, production and sales costs increasingly emerge. *"Global sourcing"* is the name for the efficient use of worldwide resources in material, personnel, capital, and energy. In short, the world is becoming a bazaar where broad competition dominates and where producers and consumers alike can shop where it is cheapest for them.

The globalization of markets and the individualization of demand are also leading to more complex manufacturing methods and specialized means of production. Universal machines are becoming increasingly unprofitable due to high set-up times and low productivity, and are being replaces by expensive special machines which necessitate a high degree of utilization. At all production stages in the automotive industry at the moment, a further concentration and reduction of the vertical integration, a lengthening of the added value chain, and thus a deepening of the division of labour and specialization can be seen. This is causing an ever greater specialization among the links in the added value chain. Those who cannot afford this specialization must withdraw.

---

[19] Conf. Ernst&Young (2004a).

Low vertical integration with many sub-suppliers requires an overall greater logistic effort on the part of all manufacturers and operators. The trend to the internationalization of markets will intensify this development. The demands on development and construction departments no longer show a linear, but a progressive growth rate in the complexity of production and ever shorter innovation cycles.[20]

Securing the future by a permanent search for cost reduction potentials is of more importance than ever for automobile companies. The ever increasing cost pressure in the development, production and marketing of new production series or model variants demands the continuous reduction of process costs. In order to reach this target, the degree of automation in production and logistics is on the one hand increased, and on the other hand the attempt is made to guarantee the production supply with ever decreasing reserve stock on site.

With the goal of keeping storage costs to a minimum, automobile manufacturers today have even important parts and assemblies delivered to the production line directly, *just-in-time* or *just-in-sequence*. Previously stocks of less cost-intensive articles were kept for one whole shift at the production site as a buffer, in order to guarantee a trouble-free production supply. With new production series, however, planning is based on a safeguard stock of only half a shift. The development of these potentials is only made possible through the application of modern communications and IT technologies.

Furthermore, development times, flexibility and customer orientation are increasingly gaining in importance. But a high degree of complexity stands in the way of these success factors. Nonetheless, many automobile organizations are characterized by highly complex product and offer programs.

The causes of this degree of complexity are to be found above all in the offer program, but also in the internal organization of the added value process. In order to gain a bigger market share, products are offered in ever more variants, to meet customer demands despite small unit figures. This all raises the costs and in borderline cases the losses, too, and decreases any tendency towards competitiveness.

Any great orientation towards exportation leads to a further increase in complexity, as the specific laws and regulations of the countries involved must be taken into account. Lacking internal standardization and develop-

---

[20] Conf. Becker H. (2001).

ment coordination between the individual sections of a company also increase complexity/costs and thus decrease competitiveness. This is because too much complexity causes, among other things, a high number of variants, high internal handling expenses and a tendency to wrong decisions and quality faults. This is the reason why the control of complexity management is one of the most important core tasks and competitive factors in the automotive industry.

It is the Japanese and Korean manufacturers who have made considerable progress in this field in the past 20 years.

2.1.3 Asian OEMs in catching up competition on quality / performance / styling

Japanese OEMs have always shown a convincingly high level of quality and reliability in their products. Above all Toyota, with the development of its own Toyota Production System (TPS), a level of efficiency and quality that obviously neither European nor American manufacturers have so far been able to reach, for all their efforts. Hyundai, the only remaining independent Korean manufacturer, is making considerable progress and has overtaken established European brands by a long way in the last few years in the fields of quality and customer satisfaction.

The past has shown that the Japanese manufacturers have been able to transfer the high quality and productivity achieved by the slimming and flexiblization of production processes to their American and European markets. In a Europe-wide ranking index[21] all four Japanese production sites surveyed were among the top eight of a total of 44 plants surveyed.

- In its Micra plant in Sunderland (GB) Nissan achieved a productivity of 99 vehicles per worker and thus first place in the European league table for the second time in a row.

- The other Japanese works, too – Toyota in Valenciennes (F) and Burnaston (GB) and Honda in Swindon (GB) – achieved on average a value of 87.5 vehicles per worker, and therefore significantly more than the European average of only 61 units per employee.

---

[21] World Market Research Center: European Automobile Productivity Index (2003).

This high level of production efficiency is paired with very high quality and reliability, as shown by the Japanese brands' regular prime positions in breakdown statistics and customer satisfaction surveys (see Table 11).

Besides efficient production and high-quality vehicles however, it is of prime importance for the success of the Japanese manufacturers that their products are in line with the European market, above all in design and equipment, and that they do justice to the tastes of their customers here.

Whilst Japanese cars until now only stood for "value for money" and reliability in Europe, Japanese OEMs are now getting into emotions and design in a bigger way. The Japanese manufacturers have so far lacked the more emotionally characterized brand image for which European consumers are prepared to pay higher prices. Only now are they beginning to translate into action the idea that competitive design is the key to success in the European market, and not only high quality at a low price. In consistence with this, Toyota, Nissan and Honda are complementing their European production sites with their own R&D departments, to be in a position to introduce more European-style models to the market (Table 10). Toyota's entry into Formula 1 racing can also be seen from an image building-point of view.

Table 10. Japanese R&D-centers in Europe (status quo 2004)

Manufacturer	Company	Headquarters, Division Offices	Employees	Current Functions
UNITED KINGDOM				
1 Honda	Honda R&D Europe (UK) Ltd.	Swindon, UK	70	1,2,3,5,6,7
2 Nissan	Nissan Design Europe Ltd.	London, UK	60	4
3 Nissan	Nissan Technical Center Europe Ltd.	Cranfield, UK	393	1,2,3,5,6,7
GERMANY				
4 Honda	Honda R&D Europe Germany Ltd.	Offenbach, Germany	90	3,4,6,8
5 Isuzu	Isuzu Motor Germany Ltd.	Gustavsburg, Germany	125	1,2,5
6 Mazda	Mazda Motor Europe G.m.b.H.	Leverkusen, Germany	231	3,4,6,7,8
7 Mitsubishi	Mitsubishi Motors R&D Europe Ltd.	Trebur, Germany	81	1,2,3,4,5,6
8 Toyota	Toyota Motor Sports Germany Ltd.	Cologne, Germany	600	9
9 Subaru	Subaru Test & Development Center (STCE)	Ingelheim on the Rhine, Germany	7	2,3
THE NETHERLANDS				
10 Mitsubishi	Mitsubishi Motors Europe B.V.	Amsterdam, The Netherlands	411	8
SPAIN				
11 Nissan	Nissan Technical Center Europe (Spain) S.A.	Barcelona, Spain	197	1,2,3,4,5,6,7
12 Subaru	Subaru Europe N.V./S.A. –Barcelona Office	Barcelona, Spain	2	6,8
FRANCE				
13 Toyota	Toyota Europe Design Development S.A.R.L. (ED2)	Montpellier, France	32	4,5,6,7,8
BELGIUM				
14 Nissan	Nissan Technical Center Europe Ltd. Brussels Branch	Louvain-la-Neuve, Belgium	16	2,3
UNITED KINGDOM				
BELGIUM				
GERMANY				
15 Toyota	N.V. Toyota Motor Europe, Marketing & Engineering S.A. (Technical Center)	Burnaston, UK Zaventem, Belgium Kerpen, Germany	233	1,2,3,5

1. Technical support for procurement of parts for local production
2. Evaluation of parts;
3. Evaluation of vehicles;
4. Styling and general design
5. Parts design;
6. Vehicle design;
7. Prototype production;
8. Marketing research
9. Development of Formula 1 racing cars;

source: JAMA

Above all Toyota has meanwhile, in the course of its strategy for conquering the European market, focussed increasingly on a "Europeanization" of its models and founded its own design center for Europe in Nice in 2001. Nissan followed this example in 2003 with the foundation of its "European Design Headquarters" in Paddington (UK), with currently 60 design employees.

"I am extremely careful about design in Europe, people form an opinion of the car at one glance, so we have got to get it right", confirmed Toyota's Europe manager Shuhei Toyoda this strategy[22]. Thus the models Yaris, Corolla compact car and Avensis Sedan were designed and produced entirely in Europe and extremely successfully introduced to the market.

2.1.4   "First we take Manhattan, then we take Berlin!" - Asian brands on the advance

Japanese automobile manufacturers have been aiming at a global market presence in all sales markets since the 70's. They were increasingly successful, at first in the USA and then in Western Europe. They were only able to reach their current position in the global market by expanding their activities in these big volume markets and by gaining market share at the cost of established OEMs.

The Japanese manufacturers were able to make continuous growth in North America and with over 30% market share they have meanwhile reached an excellent market position. This market conquest was completely at the cost of the American producers. In the meantime, American resistance to further market share losses is increasing considerably; DaimlerChrysler is playing a leading role here.

It cannot be overlooked that western and central Europe are being increasingly focussed on by the Japanese, since these have already successfully tapped into the NAFTA-zone. So far the Japanese manufacturers have had a much harder time in Europe for many reasons and their market share was stagnated at under 12% in the 90's and at the beginning of this decade.

Meanwhile though, several things point to the fact that all Japanese manufacturers want to expand their market position in Europe signifi-

---

[22] DET-News (2003-09-16.)

cantly, too, after having come through the crisis in their home market, and are looking to conquer the market in a way comparable to in the USA.

The spearhead of this movement is definitely Toyota, who was able to make considerable progress in expanding their European business in the last few years, despite the general lull in the market. But Honda and Mazda have meanwhile also registered double-digit growth rates in the European market. Nissan, too, was able last year to achieve a growth in new registrations for the first time again, only Mitsubishi had to accept sales volume losses in Europe.

The market success of the Japanese manufacturers in Europe is completely at the cost of the established European manufacturers, as the Western European automobile market was declining in the past few years. Under these difficult conditions with stagnating real incomes and cautious consumer behaviour the Japanese producers obviously see advantages for their policy of conquest. It is particularly in difficult economic times that criteria such as reliability, quality, low repair susceptibility, which are often attested to in the Japanese vehicles, gain considerable importance. These attributes, combined with an increased turning towards European design, innovative propulsion concepts and a growing choice of diesel models, provide the ideal prerequisites for further gains in market share. The current evaluation of 43,000 drivers' assessments of the quality of purchased models (Table 11) serves as confirmation of the success of Japanese manufacturers. Nine models in the top-ten list are Japanese!

Table 11. Top 10 in brand and quality evaluation

	brand			quality	
	model	*on average place*		*model*	*grade*
1	Mercedes	1.25	1	Toyota Avensis	1.15
2	BMW	2.00	2	Toyota Yaris	1.18
3	Toyota	3.00	3	Mazda Premacy	1.23
4	Porsche	4.00	4	Toyota Corolla	1.25
5	Audi	4.50	5	Toyota RAV 4	1.25
6	VW	6.00	6	Mitsubishi Colt	1.26
7	Honda	7.75	7	BMW Mini	1.30
8	Peugeot	9.50	8	Mazda MX-5	1.31
9	Škoda	9.75	9	Honda Civic	1.36
10	Mazda	10.00	10	Mazda 626	1.35

according to ADAC-AutoMarxX. date: February 2005.source: ADACmotorwelt,

The conquering of the European market by the Japanese producers is analogous to the way they did it in the USA, increasingly using their own European production sites. Above all Toyota have greatly expanded their production in Europe – within 10 years from only 12,000 units in 1992 to about 500,000 in 2003 (including Turkey, compare Fig. 27). According to Toyota manager Tokuichi Uranishi in an interview with the SZ[23], Toyota wants to achieve sales of 1.2 million cars and a market share of seven to eight per cent in Europe by 2010. They are targeting a market share which is higher than that which the European manufacturers hold outside their home countries.

In the meantime, Toyota manufactures almost as many cars in Europe as Nissan, who began as early as 1986 to build up production capacities in Sunderland (UK).

Fig. 27. Toyota passenger car production in Europe

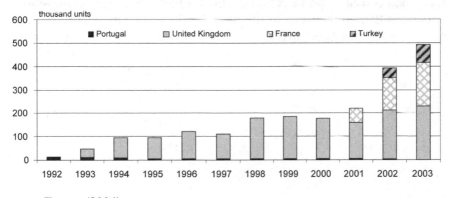

source. Toyota (2004)

Whilst Mazda have no production sites at all in Europe and saw their share in the European market halved in the 90's, Honda were at least able to make profit from their strong expansion of European production in the last two years and to increase their market share.

It must be recorded that the increasing manufacturing activities of the Japanese and also the Korean producers are not a result of increasing regional demand, but rather follow a conscious strategy to adapt better to

---

[23] Conf. SZ (2005-03-21).

prevailing market conditions and varying customer expectations, and to increase their share in the European market. Even if it is mainly producers such as Fiat or Renault who are so far affected, the examples of Opel and Volkswagen, whose main sales market is Europe, show that German OEMs are also suffering under this intensification of competition. These in turn put the other German competitors under pressure, so that the chain competition comes into full effect.

In future it will be above all the successful combination of Japanese quality and productivity with European design and sporting image which will cause increasing sales problems for European manufacturers. At the moment it is mainly the European mass producers such as Renault, PSA, Fiat, Opel, Ford, and VW who are under attack from the Japanese concerns. However, it is a foregone conclusion that Toyota will be concentrating with the *Lexus* brand on advancing into the luxury segment of the European market. In Europe today 25,000 cars of this brand are sold and in the next few years Toyota aims to treble that figure. The target set by Toyota manager Tokuichi Uranishi is 100,000! [24] In the USA Toyota have already had considerable success since the mid-80's, but at greater cost to the American than to the German suppliers BMW and DaimlerChrysler. Until now, Europe has been spared for the most part; the acid test is still to come.

---

[24] Conf. SZ (2005-03-21).

2.2   Home-made load factors

2.2.1   Boundless model palettes

Dampened growth expectations on the established volume markets, the advance of the Asian manufacturers and constantly increasing cost pressures have led to a situation in which all the manufacturers are forced to seek their "competitional salvation" in the increased differentiation of their model ranges.

- Mass producers such as Volkswagen, PSA or Toyota are pushing forward into the luxury segments.

- Luxury suppliers such as Audi, BMW and DaimlerChrysler are entering the mass markets.

- Niche suppliers such as Porsche are entering new ground and other niches with off-road vehicles.

All OEMs are therefore developing into full-line suppliers, in order to bind old customers to the expanded choice offered by the company by means of a comprehensive range of models, and at the same time to woo new customers away from the competition. To increase or at least maintain the group's market share is the top strategic target, whatever the price.

In order for a company to reach the goal of having its own comprehensive production range as soon as possible, considerable acquisitions were made. With the exception of Toyota, Honda and Porsche there is not one automobile company in the world which has not been involved in mergers with, takeovers of or share investments in other companies at some stage in the past. As a rule these involved first high financial and personnel integration costs and later, just the same "disintegration" costs, as for example with BMW-Rover, DaimlerChrysler-Mitsubishi, GM-Fiat, etc.

Obviously the failed acquisitions have on the one hand a deterrent effect, on the other hand they have also promoted the learning process within the branch. Today cooperations in specific fields of performance, such as bodywork or motor construction, or temporary production co-ops, are increasingly replacing direct takeovers of brands and competitors. Or models are designed and produced entirely by suppliers, such as the BMW X-3 at

Magna (Austria) and to a certain extent finished products are purchased externally.

This is not the place to examine the reasons for failed mergers, they are complex and sometimes elude rational analysis. More interesting are the reasons why such takeovers fitted into the strategic calculations of the companies involved in the first place. In the fore was always the strengthening of strategic market position in the crowding-out competition. Additionally it was about obtaining economies of scale in production, purchasing and distribution, which would lead to considerable cost reductions or yield growth.

Always in focus was the *strategic competitive advantage of a volume strategy*. Because size makes it easier to use platforms which, when the same components are used in different vehicles and also in different brands, make a distinct reduction in development and procurement costs possible. Small producers (i.e. Saab, Volvo, Jaguar) moreover, profit from the know-how and cost synergies which an internationally active automobile conglomerate can offer them. None of them alone has the necessary size to be able to develop and integrate new materials, ambitious electronics, complex safety systems or competitive production processes, which are the basis of modern automobile production (see at 2.1.2.2.).

Thus in the global automotive industry a mutually accelerating spiral of globalization on the one hand and concentration on the other has been developing (Fig. 28). National groups (i.e. Fiat in Italy) have been integrating smaller domestic brands (Alfa Romeo, Lancia, Maserati, Ferrari); international groups have acquired national car makers (i.e. Ford / Jaguar, Volvo; GM /Saab); and national companies have been joining up on a national level (Hyundai / Kia; Peugeot / Citroen); or internationally (VW / Audi, Škoda, Seat; Daimler / Chrysler; Renault / Nissan).

Thus the number of legally independent automobile manufacturers has shrunk continuously in the course of this concentration process. While in 1960 there were still 62 independent automobile producers worldwide, there are now only 11 (Fig. 28). The process of crowding-out and concentration in the market is now very advanced, but not yet finished. The question of which and how many producers will remain in future, or which are currently in the best position to do so, will be analysed in detail in chapter 4. It is the opinion of the IWK that the number of independent OEMs will shrink to 7 – 9 by 2015.

Fig. 28. Concentration of automobile manufacturers 1970-2015

1970	1980	1990	2004	2015
Abarth				
Alfa Romeo				
Alpin				
AMC				
Aston-Martin				
BLMC				
BMW	Alfa Romeo			
Chrysler	AMC			
Citroën	Aston-Martin			
Daimler-Benz	BL			
De Tomaso	BMW			
Fiat	Chrysler			
Ford	Daimler-			
Fuji H.I.	de Tomaso			
GM	Fiat			
Honda	Ford	BMW		
Innocenti	Fuji H.I.	Chrysler		
Isuzu	GM	Daewoo		
Lamborghini	Honda	Daimler-		
Lotus	Isuzu	Fiat		
Maserati	Lamborghini	Ford		
Mazda	Lotus	GM		
Mitsubishi	Mazda	Honda		
Nissan	Mitsubishi	Hyundai		
Peugeot	Nissan	Isuzu		
Porsche	PSA	Mitsubishi	*BMW*	
Prince	Porsche	Nissan	*Daimler-*	
Renault	Renault	PSA	*Ford*	
Rolls-Royce	Rolls-Royce	Porsche	*GM*	
Saab	Saab	Renault	*Honda*	
Seat	Seat	Rolls-Royce	*Hyundai/Kia*	
Simca/Chrysl	Suzuki	Rover	*PSA*	
Suzuki	Talbot/Matra	Suzuki	*Ren-*	?
Toyota	Toyota	Toyota	*Fiat*	
Volvo	Volvo	Volvo	*Toyota*	
VW	VW	VW	*VW*	
Σ =36	Σ =30	Σ =21	Σ =11	Σ = ?
1970	1980	1990	2004	2015

source: automobile production, IWK presentation & forecast

[25] See chapter 4.

Nevertheless the *number of brands* will continue to increase – and thereby automatically the crowding-out competition, too. For the wave of concentration, with so many acquisitions and mergers did not at all mean a corresponding decline in brand and model variety. On the contrary, the goals of the acquiring companies were to enrich their assortments with as many brands as possible and to extend their market portfolios. The result is that a more and more differentiated and extended supply is struggling for a share in a market where demand growth is slowing down more and more.

In the past the takeover of a manufacturer served in no way to buy manufacturing capacity for the production of a company's own brands, or to shut them down. It served rather to strengthen the market position of the group itself, as a whole. The brands of the bought company continue to be produced, and integrated into the assortment of the buyer company.

By preserving individual brands the OEMs are trying to cover the entire market demand as far as possible with their own organization. The independent positioning of individual brands is sometimes taken so far that the customer buying a new car often does not know to which conglomerate his brand belongs, and how many identical parts are hidden under the sheet metal of his "brand". That may certainly be in the interest of the OEMs, as brand loyalty has a decisive influence on car purchases. The customer has certain ideas and expectations, which must in his opinion be met by a certain brand. This building up of independent brand image is a protracted process which costs a lot of time and money. The mother company naturally does not want to risk this asset after a takeover. The fact that products which are almost the same can by all means be marketed at completely different image and price levels is shown by the successful marketing of the identical cross-country vehicle Touareg, by Volkswagen and Cayenne, by Porsche.

The competitive spiral takes an additional turn as a result of the fact that all producers meanwhile show sufficient entrepreneurial potential to create new niches and no longer to be dependent on the (high risk) complementary purchase of external brands. All OEMs are busily occupied with looking for new niches and developing and introducing models for them, in order to be able to react to all customer expectations as individually as possible. The number of these so-called "cross-over" models has now become difficult to keep track of, as for example, four-door coupés, cross-country vehicles of all kinds, or convertibles, etc. Also, all producers are trying to build up a so-called "full-line" assortment, that is, from the small car, through the middle range right up to the luxurious top class, to cover the whole spectrum of demand with as many models and model variants as possible.

As the explosion of choice in the luxurious top-class segment (Bentley, Maybach, Phaeton, Rolls Royce) shows, this supply behaviour leads, in the case of a limited world market, to unplanned low sales figures, high costs and – as a rule – losses and the replacement of the sales managers responsible.

2.2.2 Exploding development costs, decreasing amounts of coverage

The automotive industry faces a competitive dilemma. On the one hand all OEMs consider – analogous to the model offensives – innovations in products and processes to be indispensable for competition reasons. On the other hand, innovations are becoming more and more "marginal", that is, small advances require high and progressively increasing investment / expenditure. To put it another way: technical progress is becoming more difficult and more costly.

For a long time it has been the component suppliers who have been responsible for innovations. The OEMs demand on the suppliers for newer and newer innovations, and the speed at which they are being introduced, put all the companies involved under enormous cost pressure. For it is becoming apparent that the buyers, and not only those in the lower segments of the market, are increasingly less willing (and in a position) to pay for more comfort and security.

The core question is why OEMs have to introduce new technologies and innovations, considering the high investments? Would it not be strategically cleverer to do without and make the existing range cheaper by optimizing processes and restructuring purchasing, thereby gaining an advantage in the market?

Here we see once again the competition dilemma of the manufacturers. Such a strategy of risk aversion could indeed lead to short-term price and market advantages, because of the connected productivity increase and cost efficiency. But there is a big danger here of missing important technological developments, missing the train altogether, and thus jeopardizing the long-term success of the company.

The dilemma is that in free-market systems no enterprise – apart from monopolists – in an open market may escape new technologies and trends. And companies can even gain decisive experience from intermediate steps on the way to a new technology, which can help them to gain market advantages in the long-term.

How high the pressure on the manufacturers to keep up with certain innovations at any price really is, is demonstrated by the results of a study by McKinsey[26] (Fig. 29). According to the study certain very utility-value-oriented must-have-technologies have achieved a very high degree of market penetration in all market segments within a short time. Safety systems such as ABS, airbags, ESP, or seat-belt pretensioners assert themselves on the market within only one product generation, because of legal requirements or customer demand. Highly-priced nice-to-have technologies, above all comfort technologies such as air-conditioning, seat heating or central locking systems, tend to assert themselves well in the upper segments of the market, but take about two product life cycles to penetrate the market in depth.

Fig. 29. Introduction speed of innovations

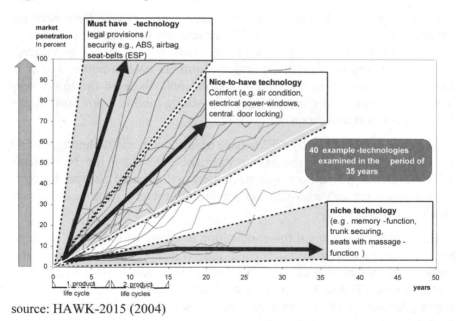

source: HAWK-2015 (2004)

---

[26] Conf. HAWK-2015 (2004).

Both of these innovation categories have a lasting influence on the OEMs' market position and competitiveness. This is because the demand for both is exclusively linked to the enthusiasm of their users or to the requirement profile of a small customer segment. By contrast, the influence of niche technologies is negligible. Even decades after its introduction, for example, the memory function for electrically adjustable seats has only achieved a low diffusion rate.

It is the markets, and legal requirements, which decide in the end which innovations take hold quickly, as must-haves or nice-to-haves, and put the companies under a collective innovation pressure. In the meantime the signs are increasing that certain customer capacity thresholds have been reached or are not far off. The German manufacturers, who, supported by an incredibly creative and technologically efficient component industry, have always presented themselves internationally as innovation pioneers, are particularly affected by such a technological countermovement. As soon as actual or putative product innovations no longer "earn" their high costs in the market, product characteristics such as high quality and reliability gain the favour of the customers, as they can be offered at lower prices by the competition. Thus it loses a substantial part of its international competitiveness.

A way out of the dilemma between

- Competition-driven compulsion to ever new technological innovations in the fields of security and comfort with less and less marginal utility for the customer on the one hand - and

- progressive upward pressure on costs for these innovations on the other hand

is the possibility of compensating the higher expenses for competition relevant innovations at another link of the added value chain. But it is not enough to adjust a few screws in the added value chain. A comprehensive new orientation of the enterprise is necessary.

In the past, operative excellence was the decisive factor in competition, leading to profitable growth. It can be seen in high employee productivity or excellent delivery quality. This operative efficiency must be improved further in the future. It is a basic requirement for successful participation in the competition, but alone no longer sufficient. Rather, a structural change in the added value chain is necessary. That is the only way to set the synergies free that make investments in innovations possible in the face of narrow competitive scope.

Fig. 30. Productivity pincer

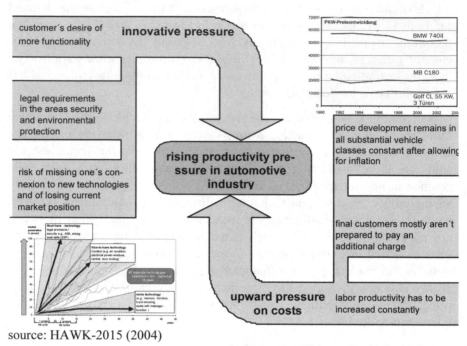

source: HAWK-2015 (2004)

This is decisive, because new technologies do not only drive competition, but are also a driving force behind increases in productivity. They lead to process optimizations, which brings direct cost advantages. In particular, they set off structural changes in the whole industry. The functionally driven added value chain develops into a piece of value-creation architecture determined by know-how. The expected synergy effects make productivity advantages of up to 20% possible. These new synergies, gained from the innovations, are breaking through the previous boundaries of the added value chain which was sorted according to vehicle modules; electronics and software are growing in significance. Traditional competitional advantages are disappearing and necessitating the building and extension of new kinds of competence. This is why the experts think the automotive industry is about to experience a further upheaval, the *innovation revolution*. [27] Only the big players of the branch – Manufacturers and component suppliers alike - will be able to afford innovations, because of progressive costs. This gives additional impetus to concentration.

---

[27] Conf. HAWK-2015 (2004)

2.2.3   Lacking model flexibility and decreasing capacity utilization

The automotive industry is well-known for being very capital intensive, because pressing plants, fully automated shell construction, paint lines, etc, require high investments. This is why profitability is a basic requirement for successful car makers. In order to attain risk-adequate profitability production capacities have to be utilized at a high average level. For these conditions to be realized even in times when demand fluctuates greatly during the whole life-cycle of products and when markets are highly volatile the automobile manufacturers' need for flexible production possibilities and works increases.

Fig. 31. Operative margin for European manufacturers' capacity utilization 2000-2006 (forecast according to WestLB)

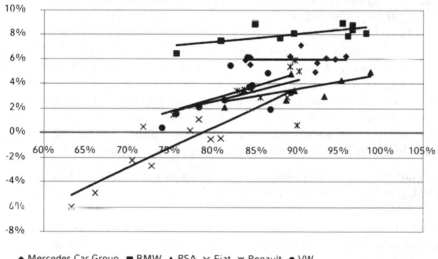

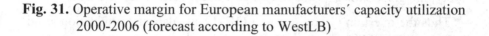

source: WestLB (2004b)

This results in a conflict between efficiency and flexibility targets in product planning. High flexibility is frequently bought at the cost of efficiency. Longer processing times, smaller system availability, higher investments, etc. are unavoidable. An efficient production structure with high variable and fixed costs will have a fundamentally steady and high capacity utilization. And then operative margins usually also high, as shown by a comparison of various European manufacturers (Fig. 31). BMW and DaimlerChrysler, as an example, have stable operative margins and the smallest fluctuations in capacity utilization. Capacity utilization at Fiat, Renault and Volkswagen is significantly smaller than at PSA. [28]

Flexibility is thus the ability to adapt to altered conditions. From the situation just described it can be concluded that in future a minimum of flexibility will be indispensable for mass producers, too, and will gain even more importance. Manufacturers, depending on their market position, will have to apply different production concepts, if they want to be able to react adequately to fluctuations in demand. As a rule, for example, a middle-of-the-range car is subject to demand fluctuations and volume changes more than a car in the luxury range. Basically, the smaller the production flexibility, the greater the expense of compensating volume fluctuations.

Thus the question of how well an OEM can adapt the production structure of its plants to fluctuating demand is relevant to competition. Here there are significant differences between the manufacturers. Studies made by WestLB[29] show that Volkswagen, for example, cannot react as quickly to volume changes as Renault. This comes from the fact that VW produces the Polo primarily in its own works in Pamplona, although an additional site in Bratislava would be available for the production of both the Seat Ibiza and the Polo. However, the site is unsuitable because it is unable to offset fluctuations during a production cycle. The Renault Clio provides a direct comparison. This model is also produced at two sites. However, a high elasticity of substitution means that fluctuations can be compensated for. This explains why Renault's flexibility is higher than that of VW.

Thus, for the OEM a main internal factor in competitiveness is its structural, that is, self controlled, flexibility in respect of product variations and volume adaptation. For automobile producers to produce more flexibly it is decisive how the individual production concepts of each plant are integrated into the whole production network of the company. The more flexible the construction in a plant, the greater the possibility of a cost-effective utilization of the manufacturing facilities, and therefore the greater its international competitiveness.

---

[28] Conf. WestLB (2004b)
[29] Conf. WestLB (2004b).

2.2.4   Crowding out competition/ overcapacities/ margin pressure

Principally, the profitability of automobile manufacturers depends on two significant factors. On the one hand this is an attractive range of products and models, because the choice of cars is still core business and the basis of success. And on the other hand it is how flexibly and efficiently the product can be produced and sold. Both factors are determined by the OEM itself, and are therefore the companies own responsibility – and not external influences.

In order to generate high profits, efficient production is a necessity, but on its own insufficient. The attractiveness of the product is decisive, and not how cost-efficiently it was produced; a fact, which does not correspond to the philosophy of all the OEMs.

Because there is a strong statistically verified connection between profitability and productivity in the automobile branch, the goal of all manufacturers is high and constant capacity utilization at lowest possible capital invested per production unit. Optimum productivity cannot be achieved in the presence of structural overcapacity or, - vice versa - in the presence of structural under-utilization. For this reason product and program planning of new models must be optimally tuned to the whole running time of the model, and perfectly adapted to the corresponding production strategy. Capacity planning which was exclusively orientated to the high initial success of a model would later involve considerable costs from under-utilization.

As mentioned above, there is a strong connection between capacity utilization and operative profit margin. Among other things, a study[30] by WestLB was able to show that a change in the utilization of capacity has a stronger effect on the operative profits of mass producers than is the case for producers in the luxury segment.

---

[30] Conf. WestLB (2004b).

Fig. 32. Capacity utilization of the biggest motor vehicle groups in 2003, in per cent

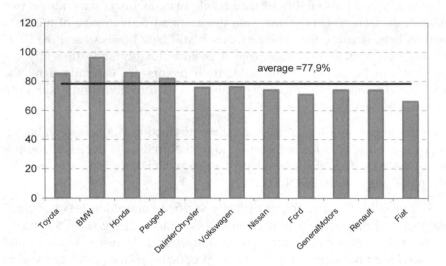

sources: PWC, WestLB, IWK calculations and presentation

It can hardly be disputed that there is a close connection between profit margin and ideal utilization. Assuming an ideal capacity of approximately 95%, then according to experts' calculations only a few OEMs can have achieved this optimum degree of utilization in 2003 (Fig. 32). According to these, the average degree of capacity utilization of the biggest car manufacturers was about 78%. It probably fell further in 2004. This low utilization of capacity reflects for the most part the difficult profit situation of the individual manufacturers.

2.2.5 Operative losses in the core segments of the "productive" added value chain/ management mistakes

Under the pressure of global destructive competition and falling profit margins on the big automobile markets, the OEMs have begun to realize that the necessary yields cannot or can hardly be made in future at the established industrial locations simply by producing vehicles. This applies to manufacturers and similarly structured component supply groups alike. A virtue has been made of this necessity, by shifting the focus of the core competence of all manufacturers from manufacturing activities towards branding and brand leadership, and above all to all kinds of downstream services to do with cars, such as financing and leasing business.

Irrespective of whether the manufacturers are withdrawing from actual car production as a matter of principle and expanding their vertical integration, in future they will have to increase the expansion of their activities in other more lucrative fields to be able to even partly maintain their profits. Here the focus is on a greater participation in follow-up business, as the profit proportion in the new cars business is considerably lower (Fig. 33).

The OEMs will probably expand their activities mainly in the following sectors:

- new and used car trade
- leasing / vehicle pool management
- financing and automobile insurance
- maintenance and accessories
- car rental business.

A number of mass manufacturers, for example, GM, Ford and Volkswagen, today no longer make most of their profits from producing cars, but from financing. Vehicle pool management and leasing have proved very profitable for manufacturers in the past. They will be expanding these fields increasingly in the future, as they can achieve much higher margins here than from the actual manufacture of vehicles.

In the fields of finance and insurance, too, the producers will extend their activities and offer more finance services and insurance to do with the product "car", with the help of their groups' own banks.

While the production of new cars is responsible for 39% of sales volume, but only contributes 8% to the profits of the manufacturers, the proportion of profits generated by financial and insurance business (46%) is much higher than that generated by sales volume.

Fig. 33. Sales volume and profit shares in the life cycle of an automobile

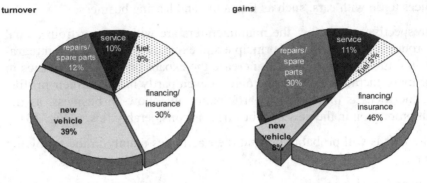

source: Booz Allen Hamilton

Meanwhile it is to be expected, that this evasion manoeuvre into "more easily worked fields" on the part of the OEMs can only bring temporary relief. Here, too, the competition will sooner or later melt the margins away.

3 Global mega trends until 2015: intensified selection process

3.1 OECD-volume markets: in saturation

In 2004 about 59 million vehicles were newly registered (passenger cars and commercial vehicles), about 5% more than in the year before. The primary market was the triad, consisting of the USA with 17.3 million (28%), Western Europe with 16.8 million (28%) and Japan with 5.8 million (10%) new cars sold. The remaining share was fallen mainly to Asia (18%), where China led the field (9 %).

Fig. 34. Regional demand structure 2004

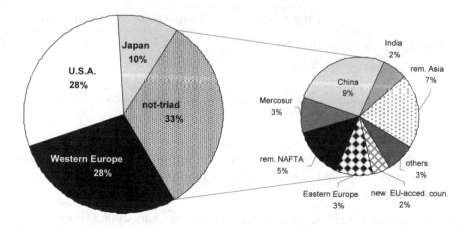

source: VDA, IWK presentation

In the past year growth in commercial vehicles (+9%) was much higher than in passenger cars (+4%) and took place mainly in Eastern Europe (+17%), Mercosur (+19%), China (+16%), and India (+25%), and therefore outside the triad. That was maybe still the sales focus for the automobile manufacturers, with over two thirds of cars sold worldwide, but the markets themselves have been showing an increasing tendency towards saturation for several years. Cyclical, that is temporary, factors alone – end of the new economy, shock of September 11, 2001, etc. – are not enough to explain this phenomenon, other forces must be at work.

It is a fact that since the 90's new registrations in the triad have been increasing more slowly than in the decades before. Since the beginning of the new millennium they are even declining in the USA and Western Europe. The actual trend goes this way in Japan, too, if one takes into account that the country has just gone through 10 years of economic crisis with a corresponding backlog of demand. Measured against this background, the slightly positive market tendency of the last few years in this country is alarmingly weak.

To conclude it can be said that the worlds highly developed volume markets are fully saturated at the beginning of the 21st century. This is a completely new situation for the manufacturers.

Table 12. Average change of new registrations in the triad, p.a.

	1980-1989	1990-1999	2000-2004
The U.S.A.	3.19%	1.89%	-0.61%
West Europe	3.50%	1.47%	-0.78%
Japan	4.24%	-1.93%	1.20%

source: VDA, IWK calculations

And things look set to stay that way. The age of steadily growing sales figures in the traditional markets of the automotive industry is obviously a thing of the past.

- The number of vehicles sold in the triad in the years 2000 through 2004 decreased on average by 0.4%. The fact that sales were declining, although producers were trying to stimulate demand with enormous sales incentives, price reductions and a multitude of new model variants, first in the USA and then increasingly in Europe, illustrates the change and the difficulties in the field and the growing competitional pressure on the car producers in their most important markets.

- The competition between OEMs for market share (safeguarding their own or gaining new shares) will probably intensify in future, because even automobile experts no longer expect a long-term, that is, trend-wise, increase in new registrations in the triad. At the New Year press conference, 2005, Professor Gottschalk of the VDA commented: „We will have to fight in the crowding-out competition for our increasing market share, with product advantages rather than with high discounts."[31]

- Later, when the economic situation improves, the backlog demand will make itself felt and lead into strong recovery phases. In the short-term, i.e. cyclically, this will sometimes look better in future, namely when customers have previously been reticent to buy new cars due to an unfavourable economy, and have postponed necessary replacements.

- However, the German automotive industry has been waiting for this for years already, without success. Sustained only by replacement demand, just such a cyclical purchasing wave in the German automobile market is very likely from the end of 2005/2006 onwards. But this does nothing to change the basic statement that in the long-term no great growth rates in inventories are to be expected, whether in Germany, Western Europe, the USA, or Japan. Short-term cyclical recovery phases do not serve to deny the long-term trend of saturation symptoms.

The reasons for this trend to stagnation in the traditional volume markets are easily understood. In the face of

- widely stagnating real income in the highly developed Western industrial states of the triad in the next 10-15 years[32]

---

[31] VDA (2005-01-27), p. 1.

[32] See also the study of the IWK (2003) "evaluation of international economic structure change and the thus deducible economic growth potentials".

Fig. 35. Growth of disposable real income

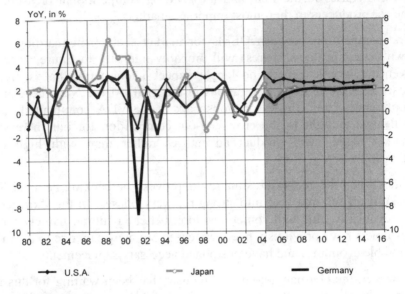

source: FERI, IWK presentation

- stagnating, or gradually shrinking, and successively ageing population in the developed world[33]

Fig. 36. Scenarios of population development

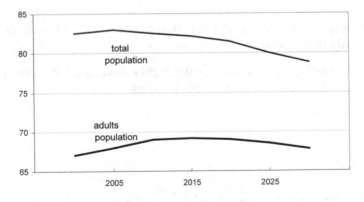

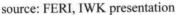

source: Shell; scenario: Tradition

---

[33] Conf. Shell (2004), p. 18.

• the degree of motorization already reached in the population of the OECD states

Fig. 37. Passenger car density by world regions

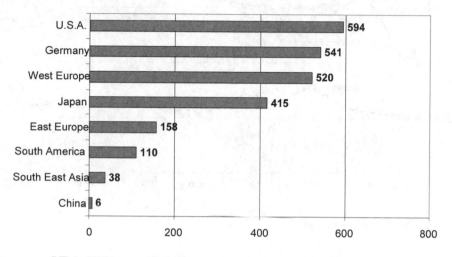

source: VDA, IWK presentation

The growth of passenger car stocks in all industrial countries will come to a halt in the course of the next 10 years. [34] New registrations, production and employment in the branch – neglecting revolutionary technical innovations which would make the global stock of cars obsolete over night – will be more and more exclusively sustained by replacement demand. In the face of the increasing life expectancy of vehicles (quality progress, declining mileage, restricted use in congested urban areas, etc.) this is postponed ever further.

Thus it becomes apparent that in the traditional volume markets of the global automotive industry is continuing to approach a saturation point at which market share gains by one manufacturer can only be made at the cost of the others, i.e. must be fought hard for (Fig. 38).

---

[34] Example Germany, conf. Shell (2004), p. 31.

Fig. 38. Forecast of new registrations in the triad until 2015, in mil.

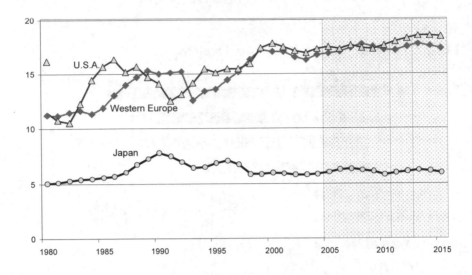

source: VDA, IWK forecast

Fig. 39. New registrations of passenger cars in Germany in accordance with inventory accession and replacement demand

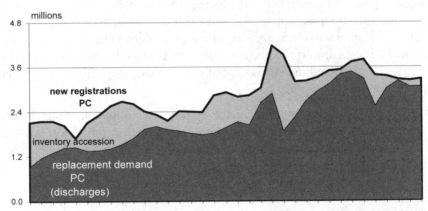

source: VDA, IWK presentation

In Germany, too, the longer-term trend will be to sinking growth rates of vehicle stock (Fig. 39), as a result of the high level already reached. Meanwhile, more than 90% of new cars sold in Germany serve only to replace old cars. Thus Shell is reckoning with a trend to sinking new registrations from 2010.[35]

3.2   Growth champions of the future: BRIC states

In the past the global economy was characterized by radical changes in economic structure and geopolitical changes of power. Among these are for example the economic rise of the USA in the 20th century, the economic miracle in Germany after World War II, with its rapid rise to being the most powerful export nation, Japan's meteoric mutation from agricultural state to industrial power, as well as the collapse of communism in the eastern block and Russia's temporary loss of power at the end of the 1980's.

In the coming decades similar global-economic changes will come from the so-called BRIC states: Brazil, Russia, India, and China. The gravitational center of international economic growth dynamics will shift clearly towards Asia, Eastern Europe and partly towards Latin America, while the USA, but above all Europe will gradually have to give up their global-economic leadership role.

Thus the 21st century will be the Asian age. Judging by all present perceptions, the development of the global economy will be characterized by the rapid advance of China and India both in the markets for industrial goods and by a very fast catching-up process – above all in China's case – in the global financial markets. Where Japan took 2 generations to develop from agricultural to industrial state, it seems that the Asian subcontinent could do it in one.

Similar developments are taking place in the central and Eastern European states, but with more modest international economic importance, compared with their economic potential. Russia and Brazil could become suppliers of raw material on a global level, China and India to producers of manufactured goods.

---

[35] Conf. Shell (2004), p. 31.

The trend increase rate in the global economy will rise on average to as much as over 4%, and thus over the long-term average of 3.7%, driven by the enormous backlog potential and the growth dynamics of these new industrial states.

If in addition effective reforms in Europe, Japan or Russia, or in other, newly industrializing countries, lead to an unleashing of market forces and increasing productivity, global economic growth could be even greater. There is at least no reason for pessimism about global growth.

Fig. 40. Contribution of the BRIC states to global growth, in %

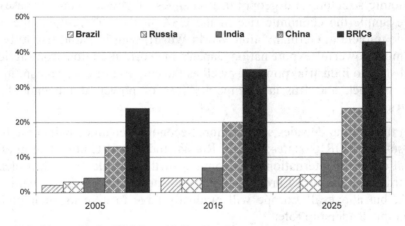

source: Goldman Sachs BRICs model projections

Other things being equal, the share of the BRIC states in world-wide economic growth will rise to almost half (40%) in the coming 20 years (Fig. 40).

Nevertheless it has to be taken into consideration that the growth dynamics of these countries are already making themselves felt, through a global shortage of important raw materials and energy sources, such as petroleum, steel, copper, etc. This means there may be certain natural limits to the expansion speed of the BRIC states. GDP growth rates in the BRIC states may well pass their peak in the next few years, but they would still – at minimum – be twice as high as in the triad states (Fig. 41).

Fig. 41. GDP forecast. Growth rates in the triad and in the BRIC countries

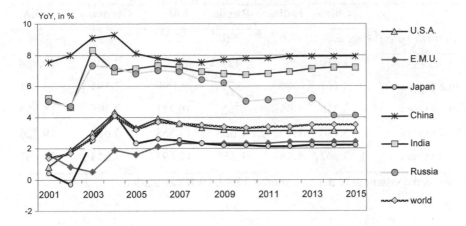

source: FERI, IWK calculations and presentation

In economic analyses the mistake is often made, of mixing up the dynamics of economic activities (say: growth rates) with the level of economic activity. If one compares the GDP level of the BRIC states, particularly China, with that of those in the triad, one can see that the world-economic significance of the BRIC states is still quite small.

This can be seen more clearly by means of the structure of world GDP. Whilst these countries together earned less than a third of the USA economy in 2003, this relation – assuming constant growth rates by all – will rise to 36% at best by 2015, and according to the Deka-Bank to two thirds of American GDP by 2050. [36]

In relation to the world economy as a whole, the weight of the BRIC countries, which in 2003 was about 8%, will increase to only a little more than 10% by 2015, according to hypothetical calculations of the Feri AG (Fig. 42), and assuming unchanged growth rates in all the countries concerned (Table 13).

---

[36] Conf. DEKA-Bank (2004), p.9.

Table 13. Growth of selected countries, GDP in billion US$

	China	India	Russia	EMU	Germany	U.S.A.
Growth rate (%)	*9,0*	*6,0*	*6,0*	*2,2*	*1,5*	*3,8*
2003	1.410	587	427	8.214	2.408	10.988
in 5 years	2.169	786	571	9.158	2.594	13.241
in 10 years	3.338	1.051	765	10.211	2.795	15.955
in 15 years	5.136	1.407	1.023	11.358	3.011	19.226
in 20 years	7.902	1.883	1.369	12.693	3.243	23.167

assumed constant growth rates; source: FERI, IWK calculations

Fig. 42. Regional distribution of world GDP

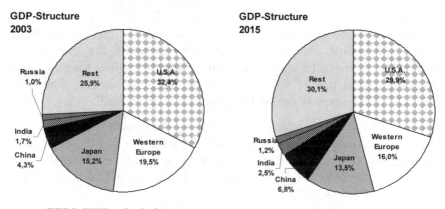

source: FERI, IWK calculations

Despite all optimism about the development of the BRIC states, there is still no reason for euphoria. For the probability that the BRIC states – as little as did the OECD states in the 20th century – are following an ideal, typical, steady growth path is not particularly strong. The risks are too high as far as the influential factors are concerned, which are in the end decisive for the growth of a national economy, as education, demographics, efficient institutions, efficiency of financial markets, the degree of trade liberalization, politics, inflation, and income distribution. According to a survey by Deka-Bank, many of these factors, above all the openness of markets and the sheer inexhaustible potential of cheap workers with high consume demand are the driving forces behind growth in these countries for the foreseeable future. But in the long-term they will lose effect mainly for demographic reasons. [37] So in the next few decades most BRIC states, above all China from 2030, will face the problems of an ageing population – just as the industrial countries today.

Big reform efforts as to the liberalization of financial markets and legal security are still necessary in the BRIC states, if the growth process is to continue at the same speed as now. Surveys by Transparency International confirm a definite connection between the degree of corruption and the economic dynamics of a country. [38] In the current evaluation of the BRIC states, they still seem to have a long way to go.

Table 14. Corruption level

country	corruption index place
Brazil	54
China	66
India	83
Russia	83
The U.S.A.	18
Germany	16

source: Corruption Perception Index, Transparency International

Regarding the future growth dynamics of the automotive industry and of the economy as a whole, one can conclude that seemingly inexorable rise of the BRIC states is hardly stoppable in the short-term, but in the long-term is with certainty not to be taken for granted. The extent, to which these countries are successful in shaping their large growth potentials from cost-effective labour reservoir, low standard of living and high consumption backlog efficiently and as a free market in the long-term, will be decisive. To this end it is necessary for the BRIC states to succeed in creating the basic conditions and prerequisites for lasting growth. For foreign investors the seriousness of their efforts, and therefore, too, the expected lastingness of the growth development identified, can be fixed to state measures in the following growth factors:

---

[37] Conf. DEKA-Bank (2004), p. 11.

[38] In that place.

- education and training,
- demography,
- efficient institutions,
- openness of markets (goods and capital markets),
- political/military stability.

3.3   New orientation of global production locations

It is an economic truism, which sooner or later offer follows demand. Since the focus of the automobile sales volume markets has shifted to the new growth regions of the world economy, vehicle production will move there, too – not over night, but little by little. The satisfaction of the automobile growth markets will take place increasingly through local production and less and less by the traditional export method, as practised successfully by the German automotive industry to date.

The consequence is that volume production at traditional European production locations such as Germany will first stagnate and then, slowly but surely, shrink. Very many small and middle-sized component suppliers are, – as shown below – at the forefront of this transformation process. Considering this, the process is taking place very unspectacularly, without the larger public noticing the gradual cutback. The revaluation of locations by investors is only noticed by the public when industrial heavyweights such as Opel, Volkswagen or Siemens, Miele, etc. threaten to close works and shift employment abroad.

This is because it is not only the new, up-and-coming markets, i.e. the developing markets which give the manufacturers a reason to build up new production locations outside the triad, or cut down old existing sites and shift them to the BRIC states. This process of migration is considerably intensified by

- the increasing competition induced cost pressure for the OEMs in their home markets, which is passed on in full to the component supply industry,

- the increase in location advantages in low-wage countries, caused by globalization and modern logistics and communication techniques,

- attractive basic economic conditions such as cheap wages, lower taxes and fewer legal regulations in the low-wage countries.

The utilization of all these cost advantages by taking up production in these countries causes an increase in price competitiveness for the manufacturers, whether OEMs or suppliers, at all levels, and in the intensely competitive domestic markets. According to experts´ estimations, it can be assumed that the production of passenger cars will rise on average by 5 % to 7% in the next few years in the BRIC states and above all in Eastern Europe. By contrast, production growth in Western Europe and the USA will tend toward zero. As related to supplier input, almost 30% of added value created in Germany in the automotive industry comes from low-wage countries, tendency rising.

Fig. 43. Passenger car production in BRIC states

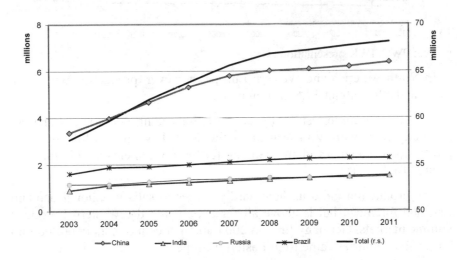

source: PricewaterhouseCoopers, Global Automobile Outlook, IWK presentation

According to Pricewaterhouse Cooper's Global Automobile Outlook the BRIC states represent about half of the passenger car production growth forecast for the next ten years. In absolute numbers, production in the BRIC states is set to rise from about 7 million to about 12 million cars between 2003 and 2011 (Fig. 43).

Fig. 44. Regional growth of passenger car production, in mil. units

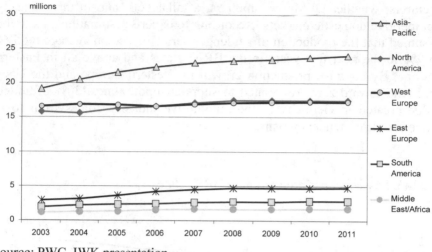

source: PWC, IWK presentation

In North America and Western Europe, however, production will come to a standstill at about 17.5 million units.

In Eastern Europe, especially in the EU-acceding countries, nearly all manufacturers, with Volkswagen in the lead, have meanwhile built up production capacities and are planning further investments here in the near future (Table 15).[39]

The production build-up here hardly serves to satisfy regional demand first and foremost, but is above all for export into the Western European volume markets, Germany first. In 2003 30.5 per cent of cars imported into the 15 EU countries came from Eastern Europe. [40]

In the short and middle-term the significance of Eastern Europe as a production location for European manufacturers, but also for those from Japan, will increase further. Average income, which is still much lower, and the lower purchasing power in Eastern European countries mean that the western manufacturers' more highly-priced vehicles are (still) too exorbitant for many people. Together with the rise in the population prosperity expected as a result of the EU enlargement, however, Eastern Europe will also gain greater significance as a market for new vehicles.

---

[39] As to the current survey about the amount of manufacturing plants of German automobile enterprises in Middle and East Europe see VDA (2004b), pp. 69- 88.
[40] Source: Eurostat

Table 15. Production locations in EU accelerating countries

OEM	Poland	Slovakia	Slovenia	Czech R.	Hungary
Fiat	car				
GM-Fiat Powertrain	engine				engine
GM/Isuzu	engine				
GM (Opel)	car				
Hyundai/Kia		car			
PSA		car			
PSA/Toyota		engine		car	
Renault			car		
Suzuki					car
Toyota	engine				
Volkswagen	car engine	car engine		car engine	car engine

car = final assembly plant   engine = engine-/gear plant

source: Automobile News Europe

Table 16. Enlargement of production capacities in Europe, per thousand units

	2003	2004	2005	2006	03-06
General Motors	-200	-150	-	-	-350
BMW	20	180	80	-	280
DaimlerChrysler	100	-	-	-	100
PSA Group	130	60	-	300	490
Nissan	100	-	-	-	100
Toyota	40	150	40	-	230
Hyundai	-	-	-	200	200
Total Europe	*190*	*240*	*120*	*500*	*1050*

2004-2006 = estimated; source: Company data & Morgan Stanley Research

In contrast to Eastern Europe, Asia is an interesting production location not only because of low wage costs, but also because of the strongly increasing local demand. Correspondingly the build-up of new production facilities in Asia is not taking place in the traditional automobile countries of Japan and South Korea, where production is stagnating in much the same way as in Western Europe and the USA, but above all in China, where the attractions are an almost inexhaustible reservoir of manpower and strong market growth.

Chinas booming economy is causing a distinct rise in incomes, at least for part of the population, in the new middle and upper classes. While in 2002 only 3.7% of the Chinese population were earning enough to be able to afford a car, by 2010 it will be 13%. In absolute figures that means a growth from 50 million potential buyers today, to more than 170 million. [41] This class of upward-movers will presumably ensure a distinct increase in sales of new cars in China.

Fig. 45. Production development of Asia

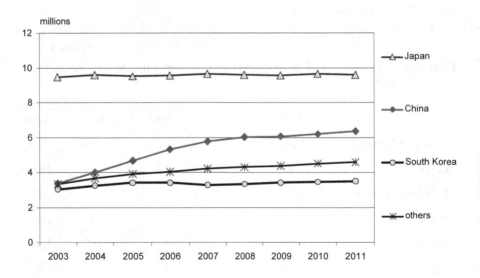

source: PWC, IWK presentation

---

[41] Conf. Mercer (2004)

Therefore almost the entire Asian production growth in the automotive industry is taking place in China. Forecast production there will increase by an average of 8.5% by 2011, at first strongly, but towards the end of the decade a temporary decline in production growth must be reckoned with. For the other Asian growth regions such as India, Thailand and Vietnam considerable production growth is also forecast, though a lot smaller than in China.

Low wage costs, paired with strongly increasing sales figures and the easier market entry conditions as a result of the Chinese government opening up for foreign trade (e.g. WTO entry in 2001), have attracted nearly all the big OEMs to China in the last few years, with their own production facilities. the number of models built in China thus records rapid growth. In the year 2000 there were fewer than 10 models produced in China by foreign manufacturers, in each of the following two years 19, and in 2003 even 38 new models were added. At the end of 2003 there were thus 60 foreign models already on the market, which were made in China (Fig. 46).

Fig. 46. Models produced in China

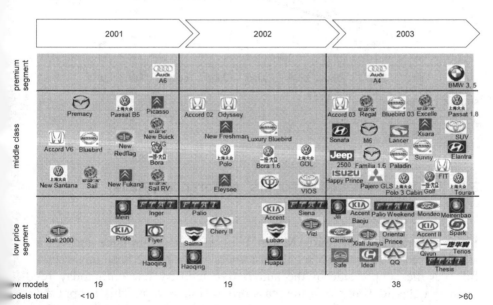

source: Mercer

Nevertheless, switching to the Chinese growth market is not a solution to earnings pressure and destructive competition for the big traditional car makers. The rapid build-up of production capacities in China exceeds the demand for new vehicles by far, which leads to overcapacities and the well-known consequences for the manufacturers of heightened competition, discounts and margin reductions. According to estimates overcapacities in China are currently almost 50% and will only decrease slowly in the next few years (see Fig. 47).

Fig. 47. Passenger car demand and production capacity in China

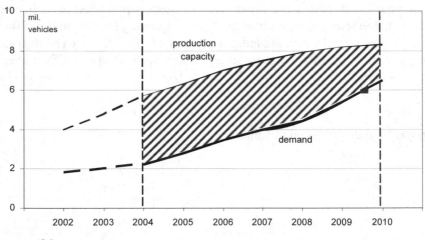

source: Mercer

3.4   Brand orientation of buyers' preferences, Asian competition advances

Since the era of Henry Ford's Model T all attempts to produce and sell a standardized world car to meet all customers' wishes have been unsuccessful. Yet the supply side of the automotive industry has until today remained a classic example of the mass production system. This is opposed by the current trend, where customer demand in saturated markets is increasingly characterized by individual wishes and clients' heterogeneous preferences.

In the last two decades the sellers' markets of the past, which were dominated by the benevolence of the suppliers, have thus developed into buyers' markets which determined by the satisfaction of very differentiated and individualized customer wishes.

In this situation all manufacturers are faced with the necessity of adapting to this change in purchasing behaviour by expanding their model ranges into niches. This is true whether they want to keep their market share in the fight for crucial customers or to conquer additional market share from competitors (Table 17).

Table 17. Share of individual vehicle segments in total German passenger car market, in %

	Traditional market segments (sedans, station wagons)	trend segments (sports cars, compact cars)	Crossover-segments (vans, SUVs/SAVs)
2010	60.5	15.5	24.0
2002	74.5	13.0	12.5
1995	87.4	8.5	4.0

source: Institute for Automobile Economy (IFA)

Thus the direction of causality in the automotive market has undergone a fundamental change. Demand no longer depends on supply; rather it is the customer who determines the variety and quality of the products supplied. Only by means of increased model variety, right up to the "full line supplier" covering the whole range of possible models and market segments, can a manufacturer be anywhere near sure of preserving customer loyalty to his brand, by catering to the comprehensive variety of changing and increasingly heterogeneous customer wishes. Besides this there is also an ever increasing number of equipment options for most models and series, so that buyers can have their cars "tailor-made" to their own taste. The whole process, from order to delivery, is moving toward total adaptation to the desires of "king client".

As a consequence of this progressive "change of power" in the automobile market, cars have increasingly become emotionally loaded consumer products for which brand image is just as important as function or the price-performance ratio. This trend will intensify in the future. To the disappointment of developers and engineers this means that branding and brand management belong increasingly to the core business of an OEM and decreasingly the actual development and production of vehicles. This is meanwhile something which can be done by large supply enterprises like Magna, for example.

For the automobile manufacturers this means that brand management is becoming a more central part of company strategy. Successful manufacturers such as BMW or Porsche are concentrating increasingly on brand-specific elements such as design, brand experience and service strategy and on product innovations, functions and technologies which form the brand profile. In the long-term it is only by creating a clear brand profile that they can differentiate their brands in the competition. The result is a distinct change in self-conception and roles in the branch. The OEMs are becoming high-tech makers of branded goods, while the suppliers are gradually taking over all the tasks involved in vehicle construction which are not brand forming[42], and which they can for many reasons perform more cheaply than the manufacturers themselves. This can be shown using BMW as an example (Fig. 48).

Fig. 48. Brand formative and non-brand formative modules with BMW as an example

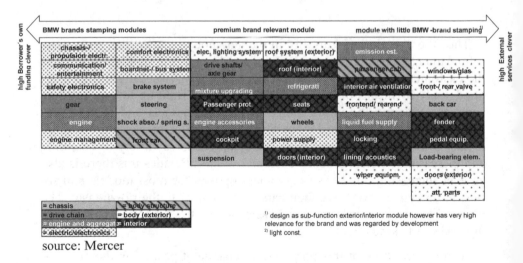

source: Mercer

---

[42] Conf. FAST-2015 (2004).

It is important that every OEM finds and determines its individual brand image, depending on its position in the lower, middle or upper segment of the market. There are no patent remedies for this, especially when more and more manufacturers are "crossing over" out of their traditional market segments – like Volkswagen into the luxury class with the Phaeton, or BMW and DaimlerChrysler into the lower market segments with the 1 series or the Smart.

To do justice to the ever more demanding customer wishes and at the same time maintain or even gain market share in the face of the competition all manufacturers are forced to expand their product ranges and improve their vehicles qualitatively and technologically. This results in a longer life-expectancy of the vehicles, while at the same time the growing fragmentation of the market leads to a true explosion of variants and types. Thus for example, the diversity of variants in automobile construction increased by more than 400 per cent in the last decade (Fig. 49). The number of special features rose by 200 per cent in the past 20 years.

Fig. 49. Development of the Mercedes-Benz model range

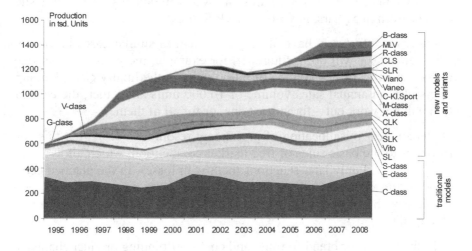

source: Mercer, Marketing Systems

The expansion of model supply can take place either by takeovers and the complementary purchase of brands (Daimler – Chrysler – Mitsubishi; BMW – Rover – Rolls Royce; Renault – Nissan; Volkswagen – Bentley –

Lamborghini – Bugatti – Škoda – Seat etc.) or by proprietary development (Toyota – Lexus- Scion).

Both strategies have advantages and disadvantages, both can be successful but may not be. It always depends on the quality of management, that is, how policy is implemented. The last 20 automobile years offer sufficient examples of both successes and spectacular failures.

Almost all manufacturers are now in the process of covering smaller market niches (cross-country vehicles, SUVs, MPVs, convertibles, roadsters, etc.) with their product lines, or are trying to create new niches, such as people carriers, 4-door coupés, hard-top convertibles, sports tourers, cross-over-models, etc. They are also trying to build up a full range, that is, to reach the whole demand spectrum with as many models as possible, from the small car and the mid-range through to the luxurious top-of-the-range.

However, all that causes considerable additional expenditure for development and logistics, which is only worthwhile if niche models can be sold in sufficient volume.

Taking the Smart as an example one could become very pessimistic. In spite of considerable sales figures of over 150,000 p.a. it has not succeeded, even in 5 years, in writing black figures.

Other manufacturers have already experienced similar cases in the luxury segment, or will do in future. If, for example, the global market only allows an annual sales volume of 5,000 units in the luxury class, but suppliers plan an annual sales volume of 10,000 units altogether, the calculation will not work out – for at least one, or for all of them.

Here a clear misallocation of investment resourced occurs, but this corresponds to the basic rules of free market competition, which allows planning errors and only later sanctions them with losses. In the planned economy system of old socialism probably none of the necessary 5,000 units would have been produced, too few, in any case – and even then of miserable quality.

In the search for brand forming and customer binding product characteristics all manufacturers are seeking their salvation in increasing the range of comfort and safety features they offer. *Product innovations* are an important contribution to differentiation. By introducing new technical features to the cars each of the manufacturers is doing its utmost to stand out against the competition.

This increases above all the pressure on suppliers to continuously develop new innovations, and at the same time strengthens their bargaining power against the OEM, if it concedes the exclusivity of these innovations. However, the time for which an innovation is exclusive is now reduced to less than one year, as the imitation competition has become more and more effective. Sarcastic people even believe that it will drop to near zero once the Chinese automotive industry reaches a certain level of technological development.

On the one hand product innovations enable sales potential to increase, on the other hand, higher margins for both supply firms and manufacturers can be achieved through the head start they give. But a decisive prerequisite is that innovations are strictly adapted to customer wishes. It does not help, to produce innovations which are technically feasible but are not demanded by the market. Thus the original two airbags became four, four became six, six became eight, etc.

But all this costs money. Money which the customer is not in every case prepared – or able – to pay for more comfort and safety. *"Nice to have"* gradually becomes *"need to have?"*.

The fact is that up to now the manufacturers could distinctly increase the equipment standard of new cars with safety features such as airbags, ABS, ESP, etc., and this was obviously in line with customer wishes. The driver-side airbag has found its way into almost all new cars within the last decade, as has the passenger airbag, and side impact and curtain airbags are also widespread. Navigation systems are now already present in about a fifth of all new cars, tendency increasing.

Traditionally innovations are first introduced in high-price vehicles, at the "upper end" of the market. For the simple reason that customers with more buying power are more likely to be in a position to, and prepared to pay the higher price of technical progress (Fig. 50). However it must be considered that the marginal costs for the new development of innovations are increasing progressively, whereas the sales figures in the high-price segments of the "old world" markets are declining. As long as the markets of the "new world" provide the necessary sales volume, that is not a problem. But if that is not so the OEMs will find themselves increasingly in a cost dilemma with their innovation policy.

Fig. 50. Readiness to pay additional charge for new technologies per vehicle class (index)

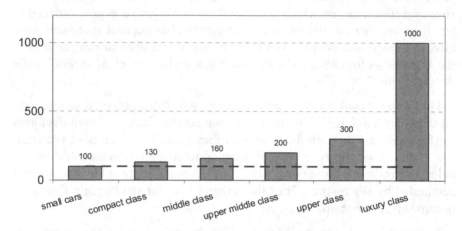

5000 final customers interviewed. source: McKinsey

Fig. 51. Equipment level in new passenger cars in Germany

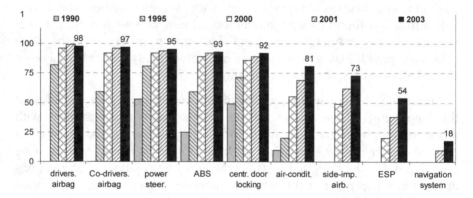

source: VDA

The entertainment and telecommunications sector offers a further field for innovations and further developments in comfort. Because of sinking average speed due to increasing traffic density, drivers spend more and more time in their cars. As a result the interior or comfort of the car gains considerably in importance. Communications systems have to be integrated here in such a way for the driver that the improved safety features mentioned above are not compromised, while passengers have unhindered access to all the possibilities of entertainment, information (internet, TV, DVD, etc.) and communication (telephone, e-mail, chat).

On the whole the trend is going toward mobile live-in offices with a strong focus on subjective wellbeing in the vehicle. Thus for example the seating comfort is ergonomically and thermo-physiologically (seat heating / cooling) better fitted to the human being and the surface feel of operating elements in the interior are gaining further significance.

Thus the car is increasingly becoming an information and communication center. Bluetooth technology and bus systems provide the networking of vehicle electronics and the driver's individual infotainment components. So the repackaging and structuring of the operator interface between driver and technical systems moves into the foreground. Moreover the importance of electronics in vehicle technology will continue to increase rapidly. Mechanical connections will be supported and replaced increasingly by electronic components. "Steer by wire" is one of many examples setting this trend.

An additional focus of technical developments is the protection of existing natural resources and the improvement of environmental protection. While in the past "end of the pipe" technologies such as filters, catalytic converters, etc were in the foreground, nowadays an integrative view of the whole process chain is taking place. Besides the further development of existing propulsion technologies and fuels to reduce emissions and protect resources, the future lies above all in the development of alternative forms of propulsion such as hydrogen, hybrid propulsion, etc. The sales successes of particle filters or the Toyota Prius with its hybrid propulsion show that these innovative technologies are also desired by customers.

With stagnating markets and increasing competition it will become more and more important for every manufacturer to differentiate itself from its competitors as a brand, and thus gain a unique selling position in the eyes of the end customer. Branding and brand leadership will become the most important task fields of the OEMs in the 21st century.

3.5 Asian competitors on the advance

As has been already described in the chapters 1.1.2 and 2.1.3, the American and European manufacturers must reckon with further gains in market share by the Asian automotive industry. The Japanese car makers, closely followed by the Koreans, are expanding their shares in all the important markets at the moment. In 2004 the share of Japanese cars in new registrations in Western Europe (just under 13%) and the USA (almost 37%) reached record levels. Accordingly, Toyota's announcement that it wants to replace General Motors as the world's biggest automobile manufacturer by 2010 must be seen as a credible company target. It is not by chance that this announcement has been made at a time when Japanese automobile manufacturers are increasing their market share against the market trend in all the world's important markets.

Besides the well-known qualities such as high reliability and a very good cost-value ratio, the improved design of Japanese and Korean cars is being increasingly honoured by Western European car buyers. Parallel to the increase in sales figures Japanese automobile manufacturers have expanded their production in Europe in the last 10 years; the European share in the global production of Japanese companies rose from a good 3% in1994 to 8% last year. By comparison, German manufacturers' share in production in Japan was exactly 0%.

According to statements made by top Japanese managers[43], the Japanese manufacturers will push their strategy of conquering the Western European market with cars from local production. In doing this they are continuing their strategy of supplying foreign markets mainly with locally produced vehicles, and not with exports from Japan. In 2001 the number of Japanese cars produced in Europe surpassed the number of Japanese exports to Europe for the first time, and is increasing distinctly, to 1.16 million vehicles (2003) compared to 0.98 Million (2003) exports from Japan to Europe. In 2004 Japanese manufacturers have already produced 1.25 million cars in Europe.

---

[43] *"We will continue our strategy, of developing, designing and constructing cars for Europeans in Europe."* (Dr. Akihiko Saito, Toyota Executive Vice President and R&D manager. In: Automobilproduktion (2004-04-07)).

Fig. 52. Japanese production in Europe and exports to Europe

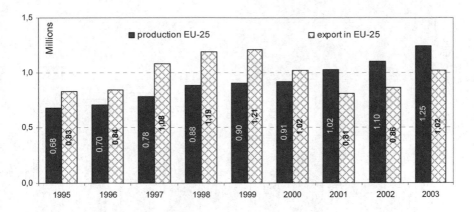

source: JAMA

On the one hand this strategy softens exchange risks. On the other hand, by this means the Japanese manufacturers automatically become members of the local industrial community; jobs are created in Europe rather than being annihilated by Japanese exports, which makes a big difference to public opinion in times of general high unemployment.

Correspondingly, components deliveries from European suppliers for Japanese manufacturers rose sharply in the past few years. In 2003 the value of "European" elements built into Japanese cars was just below 10 billion Euros, a plus of 22% on 2002. Thus a more than two-and-a-half-fold rise has been recorded since 1995. This increase can be traced back above all to the increased production by Japanese manufacturers in Europe. This is shown by the fact that the European components suppliers delivered a value share of over 80% of the parts which were used in the European production sites of the Japanese OEMs. These deliveries also include those from Japanese components suppliers which had followed "their" OEM to Europe.

Fig. 53. European parts for Japanese OEMs

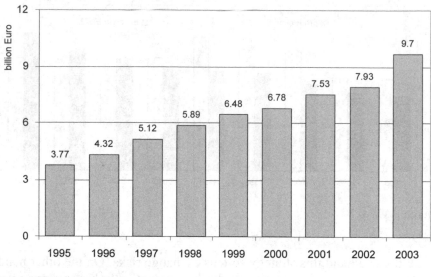

source: JAMA

It is obvious that the Japanese automotive industry has the low-cost lo-cations of Eastern Europe and Turkey as future production locations in its sights, when thinking of further production expansion. The production of Japanese cars will in future grow very much faster in Eastern Europe than in Western Europe. Production by Japanese manufacturers in Eastern Europe (including Turkey) as a proportion of their global production will roughly double by 2010 according to experts.

Japanese automobile manufacturers have realized that building up an appreciable position in a market requires local investment. This strategy was certainly the driving force for intensified activities in Europe in recent years, but also in Asia itself. Japanese companies will shift an above aver-age amount of production facilities from Japan to the Asian mainland. The homeland orientation of the Japanese has long since declined. The share of domestic production in the global production of Japanese car makers has already sunk in the last 10 years from about 76% (1994) to 66%. The IWK expects this trend to continue in the next years, and that the foreign share in total production will pass the 40% mark by 2010.

Table 18. Japanese manufacturers´ production locations in Europe (2004)

	Manufacturer Company	Location (Start of Operation)	Equity Shares	Products	Units Produced in 2003	Employees	Total Investment (million)
UNITED KINGDOM							
1	Nissan	Nissan Motor Manufacturing (UK) Ltd.	Sunderland (1986)	Nissan Europe 100%	Primera, Micra, Almera	331,924	4,45	> £ 2,000 (€ 2,941)
2	Toyota	Toyota Motor Manu-facturing (UK) Ltd. (TMUK)	Burnaston (1992) Deeside	TMEM 100%	Avensis, Corolla Engines	210,878	4,3	£ 1,700 (€ 2,500)
3	Honda	Honda of the UK Mtg. Ltd. (HUM)	Swindon (1992)	Honda Europe 86.32% Honda Motor 13.68%	Civic 3D, 5D CR-V, Engines	184,698	4	£ 1,150 (€ 1,691)
FRANCE							
4	Toyota	Toyota Motor Manufacturing France S.A.S. (TMMF)	Valenciennes (2001)	TMEM 100%	Yaris Engines	184,514	3,3	735
ITALY							
5	Daihatsu	Piaggio & C, S.p.A.	Pontedera (Pisa) (1992)	0%	Porter	7,97	5	N.A.
6	Mitsubishi	Pininfarina S.p.A. (PF)	Bairo (Torino) (1999)	MMC 0%	Pajero, Pinin	8,704	537	N.A.
SPAIN							
7	Nissan	Nissan Motor Iberica S.A.	Barcelona/Madrid (1983)	Nissan 99.74%	Terrano, Almera Tino x83 (Nissan Primastar, Renault Traffic and Opel	98,024	3,682	€ 1,625 (1990-2002)
8	Nissan	Nissan Vehiculos Industriales S.A.	Avila	NMISA 100%	Atleon, Cabster E	18,565	750	N.A.
9	Suzuki	Santana-Motor S.A.	Linares (1985)	0%	Vitara, Jimny	21,947	589	N.A.
PORTUGAL							
10	Toyota	Salvador Caetano I.M.V.T. S.A.	Ovar (1968)	TMC 27%	Dyna, Hiace Optimo	2,413	350	N.A.
11	Mitsubishi F	Mitsubishi Fuso Truck Europe S.A. (MFTE)	Tramagal	Mitsubishi Fuso 99%	Canter	5,904	300	€ 27.5
THE NETHERLANDS							
12	Mitsubishi	Netherlands Car B.V. (NedCar)	Born (1991)	MMC 85% MME 15%	Colt, Space Star, Smart forfour	163,080 (74,352)	4	NGL 4,862 (€ 2,206)
HUNGARY							
13	Suzuki	Magyar Suzuki Corporation	Esztergom (1992)	Suzuki 97,5%	Wagon R+, Ignis	88,5	2,1	N.A.
POLAND							
14	Isuzu	Isuzu Motors Polska Sp.zo.o.	Tychy (1999)	Isuzu 40%	Diesel Engines	271	1,033 Yen 26,000 (€ 194)	
15	Toyota	Toyota Motor Manufactur-ing Poland Sp.zo.o.	Walbrzych (2002)	TMEM 100%	Transmissions	134,222	1	100
16	Toyota	Toyota Motor Industries Poland Sp.zo.o. (TMIP)	Jelcz Laskowice (2005)	TMEM 60%	Diesel Engines		350 (Plan)	200
CZECH REPUBLIC							
17	Toyota	Toyota Peugeot Citroën Automobile Czech (TPCA)	Kolin (2005)	TMC 50 %	Small Passenger Cars		3,000 (Plan)	€ 1,500 (Plan)
GERMANY							
18	Mitsubishi	MDC Power GmbH	Kölleda-Kiebitzhöhe	MMC 50%	Gasoline Engines		450	2,5

source: JAMA

Toyota is a good illustration of the latest attack of the Japanese manu-facturers on the "European fortress". Although Toyota vehicles were first made under license in Europe in 1971, Toyota did not really start European production until 1992, at its works in Great Britain. In 2001 in France production of the Yaris began, which had been specifically developed in Europe. In 2002 the works in Turkey were extended as the European production base for the Corolla. This site was very much extended (from 100,000 to 150,000 units) for the new compact MPV Corolla Verso and the capacities in both plants in Great Britain and France were also increased by changing to production in three shifts. In addition Toyota invested in the Eastern European states with the building of a gear production plant in Poland and from 2005 a further production site in the Czech Republic for up to 300,000 small cars (100,000 under the Toyota name) in a joint venture with PSA, which will complement Toyota's European production. Altogether these current extensions represent an expansion of Toyota's European production to an extent of roughly 50%, and that in times when the European competition is complaining of overcapacities of up to 25%.

Toyota's strategy also includes *diversification into individual brands*. However, brands are not bought from the competition for this, although that would be easily possible with Toyota's enormous financial power (19 billion Euro cash reserve alone in 2003). Instead Toyota is banking on its own growth and the establishment of its own new brands.

For this reason sales of the in-house *premium brand Lexus* are to be pushed in Europe. 14 years after the brand's foundation Lexus is meanwhile extremely successful in the US American market, while in Europe the brand has not really been established at all. In 2004 almost 290,000 Lexus were sold in the USA, but in Germany with only 2,593 sold, less than 1 per cent of the US sales volume.

Now the brand is to be re-established with an independent Lexus design, to which end a special Lexus development center was set up last July. The marketing expenses for Europe should amount to about 500 million Euros for the next three years, three times as much as in previous years. In this way Toyota wants to increase the number of Lexus models sold in Europe to 100,000 per year by the end of this decade, more than the total number of BMW 7 series and Mercedes S class together sold in Europe today.

Because Toyota does not consider it feasible to newly develop each component part for the new Lexus models, the Japanese firm is planning to make use of synergies with its mass models. However, new and innovative technologies will be introduced first in the Lexus models. Thus before long large volume engines with hybrid propulsion and smaller diesel engines should be available for the Lexus. By this method Toyota wants to then attack the main competitors in Europe in the upper segment, Mercedes, BMW and Audi.

An introduction of Toyota's third group brand Scion, which was developed for the young target group in the USA, is not planned for Europe, at least in the middle-term. The plan is to first establish it in the American market while in Europe the Lexus brand enjoys priority. If the Scion proves successful in the USA it will only be a matter of time before it comes to Europe.

This long-term model and brand policy is typical of Toyota. Company strategy is characterized by not making precipitate or high-risk model changes or market introductions, but instead by making gradual changes in new models and testing anything new for market success in a limited region first. This means that high losses can be avoided and with patience very considerable profits made.

In the meantime a similar development to that of the Japanese manufacturers is being made by the *Korean brands*, above all Hyundai, but with a time lag of 10-20 years. At the moment the Koreans are gaining market share above all in the small car and minivan segments in Europe and the USA, because of their good price-performance ratio. In design and emotionalization of their brands the Korean manufacturers are also plainly on the advance, if not yet quite so advanced as their Japanese counterparts. In contrast to these they do not present an immediate threat for the European premium suppliers, at least for the present.

However, in the USA the Koreans now want to try to advance into the top segment with new models. If they succeed they will probably transfer their successful US strategy to the European market in the same way as the Japanese OEMs.

In the middle and long-term the established American, Japanese and European OEMs will also have to reckon with new *competition from Asia*, the Chinese OEMs. The Chinese automobile manufacturers were able to profit from the car boom in their homeland in the past few years and to study extensively and get to know the technology and production methods of western manufacturers in joint ventures. Considering this development, it is not very astonishing that the biggest Chinese car makers are now looking for other sales areas, not only in the American market, which is considered less sophisticated, but also in the European market. The high profits which they have made in the last few years in their largely insulated domestic market are responsible for this. Thus the biggest Chinese automobile producer, Shanghai Automobile (SAIC), which together with VW builds and distributes among others the models Santana, Passat and Touran in China, is for example currently in the process of taking over the British brand MG Rover, and thus a (small) VW competitor in Europe. A common enterprise is planned, of which SAIC will hold 70 per cent. MG Rover will contribute brands, technology and productions, SAIC 1.4 billion Euros. MG Rover is already the second manufacturer to have been incorporated by SAIC. In October 2004 48.9% of the South Korean manufacturers Ssangyong were acquired. Moreover SAIC is considering a takeover of the insolvent Daewoo plant in Poland.

Thus in the long-term the Chinese competition could become a menace for the European manufacturers similar to the Japanese and South Korean competitors, even if until now it was thought that a Chinese advance into Europe would be above all at the cost of other foreign brands. But it is in any case a fact that every additional competitor "in the old world" means additional price pressure and crowding-out competition, considering the saturated markets there. And Europe's manufacturers are afraid of that, for due to the overproduction in Europe cars here can only be sold with high discounts and shrinking margins. This development would definitely be intensified by additional price pressure from China.

3.6 Concentration at all levels: upheaval in the added value chain

First let me mention that the literature on expected changes and restructuring of the automobile added value chain is legion. All the consulting enterprises, most of them American, have dealt extensively with the subject. [44] As a rule, it is all about technical details such as the percentage transition of work content from OEMs to the component supply chain, right through to taxes (cent precise) and savings in overhead cost blocks of the OEMs, etc.

This may all be true, but it also may not be. Because, as is well-known, all forecasts of changes in social systems are tainted by insecurity. In this publication the expectations and extrapolations of the parties involved in the processes are always taken as the basis, which can then be extrapolated into the future as a "linear" value (today x-% share, tomorrow y-% share).

---

[44] Conf. for example the studies: Bayern Innovativ (2002), Dudenhöffer, F. (2002), Ernst&Young (2004a), FAST-2015 (2004), Fraunhofer ISI (2004), HAWK 2015 (2003), Intra (2004), Mercer (2003), Radtke, Ph. / Abele, E. / Zielke, A. (McKinsey & Company – edit.) (2004), VDA (2001), VDA (2002c) etc.

In short, there is no exact analysis of the causes of these changes, the question "why" is missing. (Einstein)

To prepare the answer to this question was the purpose of the previous chapters. While the volume of new vehicle sales in the triad no longer shows any lasting growth potential, all manufacturers are extending capacity undeterred and overcapacities are thus increasing further. Almost a quarter of European automobile production is stockpiling at the beginning of 2005, with massive pressure on earnings as a result. At the same time further billions are being invested in the building of new production locations in the newly industrializing Asian countries and especially in China, in order to be present in the only significant growth market. A problem here is that for a long time yet the best growth rates in Asia will not be able to compensate for the lacking dynamics in the traditional sales markets, because the unit figures will remain low here for the time being.

Shrinking earnings in core business force the automobile manufacturers to concentrate harder on tasks which are as profitable as possible in future. This leads to

- the *outsourcing* of all activities which can be made or performed more cost-effectively by upstream links of the added value chain (as a rule suppliers, development services, etc.) than by the manufacturer itself. this includes a shift of production to low-wage locations, which the supplier can implement without problems, but not the manufacturer, because of political constraints.

- the *insourcing* of highly profitable activities which are downstream of production, above all distribution, customer services and financial services (financing, leasing, etc.)

A number of arguments speak for concentration on this so-called "downstream" business.

- Customer contact and image are becoming decisive factors for success in brand competition. High international production standards and technology which becomes obsolete at an ever increasing rate provide an ever decreasing differentiation potential for the products. Brand differentiation is therefore increasingly sought in the brand experience.

- "Downstream" investments in distribution, services and financial services have considerably less capital lockup than investments in new technologies and production facilities and promise distinctly better equity returns, especially as capital intensity in the automotive industry is very high in comparison to other branches. Investment in real capital is increasingly left to the suppliers. The OEM's capital which is thus freed up is instead put into the expansion of financial services or the development of new models, with the result that in the long-term profits in the actual automobile business sink even further because of the competition.

- 800 million vehicles in the global market form a hitherto insufficiently tapped reservoir of business and customer binding possibilities.

The German automotive industry is at present in its most suspenseful and creative phase since the last big economic crisis at the beginning of the 90's. Innovation and technical progress have since been recognized as differentiating characteristics in competition and systematically pursued. This is seen in the considerable increase in R&D expenditure and in the many approaching new starts. In 2005 alone 150 new models will come on to the market, 80 of them have been developed by German automobile manufacturers.[45]

In this context the OEMs not only shift production responsibility, but also more and more development responsibility for new models to the suppliers. Above all these have to secure know-how in the organization and realization of series start-ups and must be able to manage complexity, besides providing ready-made product concepts.

An integrated professional start-up management which manages the internal and external partners as a network and focuses on the common goal thus becomes a decisive success factor.

---

[45] Conf. Reichle J. "Anew. Forecast: The most important cars 2005". In: Süddeutsche Zeitung no. 17 of *22nd* /*23rd* of January 2005, p. 17; as well as "Tired Start of Car industry into the new Year"., in: Süddeutsche Zeitung no. 22 of January 28th 2005, p. 14.

Fig. 54. Downstream services, related to customer and vehicle

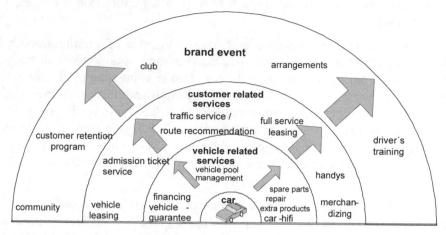

source: Mercer

Beside the constant search for excellence in operative business, automobile manufacturers have long ago left their traditional role as largely full scope producers. Only by focussing more on core competence and the better utilization of resources connected with that, on the part of both the manufacturers and the suppliers, can profit margins be stabilized in competition-intensive markets.

Most automobile manufacturers have already introduced modular strategies, which mean, that vehicles consist mainly of finished modules which various suppliers deliver ready-made to the OEMs' plants. This creates clear interfaces to the suppliers and makes flowing transitions of model and innovation cycles possible. According to expert estimates only 65% of automobile added value will in future follow the established hierarchical pattern of cooperation between automobile manufacturer and supply industry.

In order to open up these new business potentials the manufacturers will have to consistently re-align their previous business model. Thus the manufacturers development and production penetration, which has for years been declining will sink further with every model. Accordingly, suppliers will be increasingly involved in the development and production process, while the automobile manufacturers will concentrate more on brand and customer management. Within the added value chain they take over technical integration and thereby bring a network of specialist together. A part of this is that the actual production of the vehicles is increasingly transferred to suppliers of even to other manufacturers with excellent production competence in certain sub areas (gears, motors, bodywork, etc.) (see chapter 3.7).

A cooperation between Jaguar and Audi in the area of aluminium body repairs showed first signs of this. Because building up a network of body shops would not be profitable for Jaguar, Jaguar wants damaged vehicles with aluminium bodywork to be repaired by Audi and thus profit from the competence in light construction which Audi has built up in the past. Moreover, in the development of the current XJ Jaguar has fallen back upon the Alusuisse affiliate Alcan, which was already involved in the development of the spade frame architecture of the Audi A8 and the Audi A2.

More than 20 new cooperation forms can meanwhile be identified (system cooperations, production cooperations, engineering services, spin-offs or make-to-order production, etc.). By the use of specialist competence effects of scale and advantages in the distribution of risk – both in development and in production – can be exploited. The new quality of cooperation can improve EBIT margins by about 3% and return on capital by 4-10%.[46]

Fig. 55. Concentration on brand and customer management

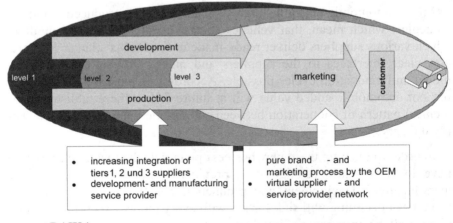

source: BAIKA

---

[46] Conf. FAST-2015 (2004).

This change will make automobile manufacturers focus more on customer oriented processes such as brand management and distribution in the future. In this way the car will become an emotionally engaged consume good. Brands will increasingly appeal to buyers on the basis of emotions. Technical performance data thus fades into the background in comparison to the total vehicle integrity, the lifestyle aspect moves into the foreground. This is underlined not least by the manufacturers' advertising.

When the automobile manufacturers concentrate increasingly on "downstream" business, both development and production in the actual core business shift increasingly into the supply industry. The suppliers take over step by step all non-brand forming tasks in vehicle construction. Today manufacturers already only develop and build their cars to 35% themselves – according to the study FAST-2015 (2004) the personal contribution currently amounts to 4,000 Euros per "average car". By 2015 it will have fallen to 2,670 Euros or 23 per cent, and the rest will be contributed by suppliers and service providers (Fig. 56). [47]

Fig. 56. Added value/added value share automobile manufacturer

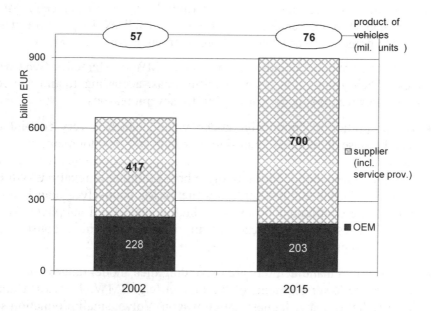

source: Automobilentwicklung

---

[47] Full particulars conf. study FAST-2015

Bodywork, sheet metal, paintwork and chassis are at the center of this development. The manufacturers will also pull further out of the construction and assembly of modules. Their advance and series development will remain fairly stable with a size of about 30 billion Euros. The automobile manufacturers will in future only invest heavily in automobile electronics. Altogether the dependency of the manufacturers on their suppliers will continue to grow. The manufacturers' own added value will come consistently from brand and product differentiation in future.

Which areas of production are sourced out is determined increasingly by the positioning of the automobile brands. 80 per cent of the top managers surveyed by Mercer[48] expect that manufacturers' own added value will come consistently from brand and product differentiation in future. The results of the Mercer survey show that the "own" added value retained by the cluster *premium / quality / sport* with brands such as Audi or BMW will be about 25 per cent higher than that retained by the cluster mass / price / comfort with brand such as Daihatsu, Kia or Rover.

The evaluation of Mercer comes to the conclusion that with few exceptions all automobile manufacturers will not only reduce their personal contribution relatively, but also in absolute value. Above all the mass producers are still reducing their own added value by up to 30 per cent. Altogether their absolute personal contribution will fall by 15 per cent. This applies to brands such as Chrysler, Ford, Citroen or Nissan.

However, premium brands such as Audi, BMW or Mercedes-Benz will increase their value addition in specific areas, according to the Mercer survey. However, this is not convincing for several reasons:

- The proportion of "innovative added value", which is very cost-intensive because it requires intensive research and development, in total value creation is continuing to increase in premium vehicles for the reasons mentioned above. In future these brand forming innovations will be contributed even less frequently by the OEMs themselves than before, but will be contributed by external highly specialized suppliers. With the X3 series BMW is already having a premium vehicle constructed solely by suppliers.

- The premium manufacturers are expanding their model ranges above all towards the lower segments of the market (e.g. BMW: 1-Series, DaimlerChrysler: B-Class, Jaguar: station wagon, Volvo: small production series, etc.). This is not only where the crowding-out competition is most

---

[48] Conf. Mercer (2003).

fierce, it is also where manoeuvring space for manufacturers' prices is smaller, due to customer power, and therefore pressure on margins is higher. The constraints on premium manufacturers to exploit and open up cost reduction potentials are very much bigger. This can only be realized by an even greater level of outsourcing.

Structurally speaking, the proportion of internally contributed added value by premium manufacturers should decline more strongly than is the case with the mass producers, which have more possibilities to shift cost-intensive parts as internal production to low-wage countries. This fits in with the fact that according to Mercer all brands must increase external contributions distinctly, in some cases to more than double the present amount.

3.7 Forming of strategic alliances at OEMs

Apart from takeovers, mergers and "amalgamations of equals", cooperations, strategic alliances, etc. have recently been gaining increasing significance for automobile manufacturers. This is surprising inasmuch as all the OEMs involved are in fierce competition with each other.

The reasons for this readiness to co-operate with fierce competitors vary from company to company, but are principally always the same, namely *cost reduction* by

- better utilization of idle capacities,
- realization of economies of scale,
- improvement of the company's own strategic market position by faster extension of the model range instead of troublesome and cost intensive me-too-products

The practical realization and the actual object of strategic cooperations can be extremely varied and take place on various levels. The most current strategic contents of such alliances and cooperation partnerships lie in the following areas:

- the exchange of aggregates and components
- common development of vehicles
- common research activities
- common production and assembly plants
- common marketing activities

Mostly these alliances are based on clearly defined cooperation contracts, however, they may be additionally supported by mutual capital interests or the foundation of joint venture enterprises. The *cooperations* can also be restricted regionally or to affiliated enterprises. Strategic alliances normally are not pursued with the target, of eventually leading to mergers or acquisitions. They are rather a stable form of working together to reduce costs or open up new markets.

Meanwhile all the big automobile manufacturers are connected to one or more of their *competitors* by strategic alliances. Examples of this are the building of common production facilities by PSA and Toyota in the Czech Republic, the joint development and production of a new generation of automatic gears by Ford and General Motors, and the cooperation between BMW and PSA for the development of gasoline engines.

The effects of this newest trend towards cooperations in the branch are shown by the latest supply agreement between Mitsubishi, PSA and Volkswagen:

- from 2007 Mitsubishi Motors (MMC) will construct sports utility vehicles (SUV) for Peugeot/Citroën (PSA), who previously were not represented in this rapidly growing market,

- the Diesel engines for these SUV will be supplied by Volkswagen, one of the fiercest competitors of PSA in the field of diesel propulsion.

With these cooperations, born of the need to control costs and the income dilemma, conglomerates or partnerships of convenience come increasingly into being between the big automobile manufacturers. The manufacturers' regional origin plays no role in this, but rather global strategic alignment and the supplementing of the model program. Thus it will not come to national, i.e., German, European, American, or Japanese conglomerates, but rather to globally aligned partnerships of convenience.

This new power constellation presents a new challenge for the component supply industry. This would be all the graver if strategic alliances similar to those in the aviation industry were to arise from these partial cooperative alliances of OEMs, for example by a division of the individual market segments between partners, common purchasing, common development, etc. That would take a lot of the heat out of the present constellation of fierce crowding-out competition between the manufacturers without leaving any possibility for fair trading offices to intervene.

3.8 Changed energy supply conditions

A decisive factor for automobile sales volume is and remains the availability of energy at affordable prices for the propulsion of the vehicles. Today 99 per cent of all vehicles are still gasoline or diesel fuelled and thus use refined petroleum. The question of how sure the supply is, is therefore of the utmost importance in determining the mega-trends in the branch, both for customers and manufacturers.

If this natural resource became scarce in future it would inevitably have an effect on the automotive industry. Potential buyers would not purchase new cars with conventional propulsion if an acute scarcity of this fuel and an ensuing rise in price were foreseeable. Alternative propulsion systems are already being developed and also produced, but do not play a role as yet in sales volume.

At the end of 2003 there were global petroleum reserves of approximately 1,137 billion barrels, of which just under two thirds were in the Middle East (compare Table 19). Measured at the current production rate the oil reserves known to the OPEC countries would last for the next 90 years, the reserves of the non-OPEC countries however, only for the next 17 years.[49]

Table 19. World-wide oil reserves by regions, 2003

	billion barrels	per cent
North America	27.2	2.4%
Latin America	116.4	10.2%
Eastern Europe	88.2	7.8%
Western Europe	18.3	1.6%
Middle East	735.8	64.7%
Africa	105.5	9.3%
Asia and Pacific	45.8	4.0%
Total world	1137.5	100.0%

source: EC Annual Statistical Bulletin 2003

---

[49] Conf. OPEC (2004), p. 5.

In a growing global economy however, oil consumption will rise, in spite of all the modern possibilities for economizing. If oil consumption were to rise at the same rate as the global economy it would be about 37 billion barrels a year higher by 2015, about 50% more than today. But because of innovations and the growing spread of modern energy saving technologies and the increasing use of alternative energy sources OPEC only expects oil consumption to rise by about 32 billion barrels by 2015. That would mean that currently known oil reserves would irrevocably run out by the middle of this century.

However, as in the past, further new oil supplies will be discovered, so that petroleum will still be the most important source of energy in 50 years' time and will still be available in sufficient quantity. Nobody can forecast today what oil reserves there still are in the world, but the past has shown that more new occurrences were discovered yearly than the amount of oil that was consumed. In 1981 known petroleum reserves amounted to 670 billion barrels at a consumption rate of 21 billion barrels. Since then known reserves have risen by a factor of 1.6 but the annual amount of crude oil produced only rose by a factor of 1.3.

Fig. 57. Growth rates of oil production and reserves (1983 = 100)

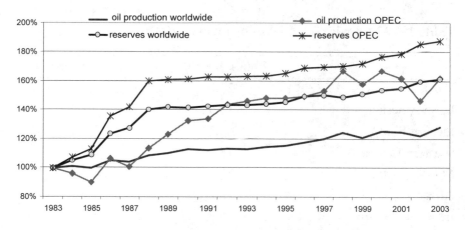

source: OPEC Annual Statistical Bulletin 2003, IWK presentation

However, the costs of exploration and above all the production costs of crude oil are becoming increasingly more expensive, as the remaining occurrences are in places which are increasingly difficult of access. These costs will have an effect on the price of crude oil and thus on *car mainte-nance costs*. However, because the crude oil price was influenced by in the past by a variety of factors (OPEC, political crises in the Middle East, etc.) and was extremely volatile, these increasing production costs will not be of significance for price development in the next 15 years.

Fig. 58. Oil price development

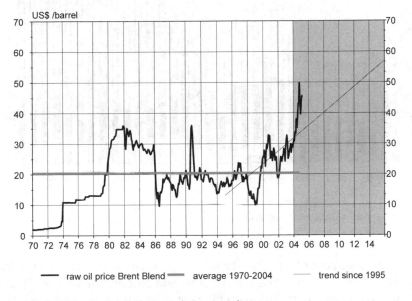

source: FERI, IWK calculations and presentation

Fiscal aspects play a more significant role however than the price of crude oil for the fuel price the end consumer has to pay. The tax portion of the gasoline price has grown continuously over years. Mineral oil tax in Germany rose by about 35 cents since the beginning of the 90's alone, and today comes to 65.5% per litre, including the eco-tax. Value-added-tax at 16% is also charged on the mineral oil tax, so that the tax portion comes to three quarters of the gasoline price: of a gasoline price of 105 cents per litre 80 cents flow into the bags of the revenue authorities. The continuous extra siphoning off of the car drivers' purchasing power through rising payments for running their cars (above all gasoline costs) has worked as an economic and psychological hindrance in the past and slowed down new car purchases and the domestic economy considerably.

Fig. 59. Composition of the fuel price (euro super) in Germany

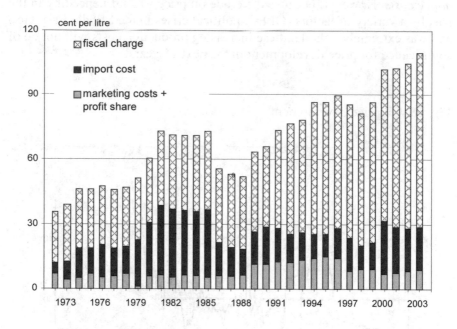

source: mineral oil economy association MWV

At the manufacturers' end the rapidly increased prices for commodities and raw materials such as oil, rubber, steel and aluminium have led to considerable unplanned extra strain on the cost side. As a trend, these will also remain high, considering the increasing demand from the global growth regions China, India and the central and Eastern European states and force down the incomes of all companies in the automotive industry. Scope for price increases which used to be available to OEMs to improve profit margins in economically more "relaxed" competition will in future be eaten away by changed factor price relations for energy and semi-manufactured products. In concrete terms this means that cars will indeed become more expensive in real terms in future and thus eat away well-funded demand, but without leading to higher profits for the OEMs.

Example

A current survey by Credit Suisse First Boston[50] for the year 2004 comes to the conclusion that in the most unfavourable case up to 1.5 billion Euros in higher costs will be incurred for Daimler-Chrysler and 1.4 billion Euro for the Volkswagen group due to increased prices for semi-manufactured products. Porsche is least affected with only 24 million Euros. Above all increase in the steel price of up to 80% effect the calculation models. Moreover a price increase of 28% must be reckoned with for plastics and 17% for aluminium.

Even in the conservative calculation model the consequences of increased prices for earnings before interest and tax are significant. BMW and Porsche have the least potential risk according to this calculation, because these companies have the widest margins. Manufacturers with the narrowest margins in the automobile business, i.e. the mass producers, are feeling the consequences more intensively.

3.9 Conclusion: large structural change in the global automotive industry

The global automotive industry is facing a heightened process of selection and concentration, beginning with the OEMs and going right through all supply levels. According to studies by Mercer and the Fraunhofer Institutes[51], their numbers will decrease from 5,500 at present to 2,800 by 2015. Of the eleven independent automobile companies probably only nine or ten[52] will then still be independent. Altogether, though, the concentration

---

[50] FAZ (2004-09-13)

[51] The study "Future Automotive industry Structure (FAST) 2015" (2004) by Mercer Management Consulting and the Fraunhofer Institutes for Production Technique and Automation (IPA) as well as for Material Flow and Logistics (IML), is based on 60 interviews with decision makers of the first and second level by automobile manufacturers, suppliers and service providers as well as an analysis of all disposable data sources and model policy of the brands.

[52] The study "Future Automotive industry Structure (FAST) 2015" (2004) by Mercer Management Consulting and the Fraunhofer Institutes for Production Tech-

processes in the branch will slow down as it approaches close oligopoly. Strategic alliances between the individual OEMs and suppliers, which lead to peaceful competitive behaviour, are contributing to this situation.

The growth regions for sales volume are above all China, India and Eastern Europe, while the highly developed markets of the USA, Japan and Europe are mostly stagnating.

The centers of automobile production are moving slowly but surely into the low-cost regions of the world (Asia and central and Eastern Europe) and thus to a greater extent into those countries where the sales markets are big, as in China. The US automotive industry will retain its superior importance, although the original American manufacturers will be increasingly replaced by Asian and European manufacturers. Europe, too, will hardly lose any of its importance as a production location for the automotive industry, although a distinct shift from west to east will take place. As a production location Japan will experience a tendential shrinkage and at best stagnate. The new works and development centers of the Japanese manufacturers are being built in Eastern Europe and America, and on the Asian mainland, above all in China of course.

Concerning the internal structure of the automotive industry itself, the following trends may be emphasized:

- Car manufacturers will become high-tech brand article providers. Brand management (design, brand experience, service strategies, functions, and technologies), image, marketing, services, and customer contact and care will become the decisive factors in competition. High international quality standards, rapid imitation competition and the ever-increasing rate at which technology becomes obsolete will mean faster-shrinking differentiation potentials for products.

- The premium brands will become the "mission statements" of their companies. This is where the core competence is, and it is where junior management staff is trained. Techniques and know-how flow from the premium brands to the mass brands. Every brand will need a clear strategy for value creation in future, in which their personal contribution profile and the necessary competence capacity and partnerships are laid down.

---

nique and Automatisation (IPA) as well as for Material Flow and Logistics (IML), is based on 60 interviews with decision makers of the first and second level by automobile manufacturers, suppliers and service providers as well as an analysis of all disposable data sources and model policy of the brands.

- The mass brands will reduce their value creation more than the premium brands, which need more exclusive characteristics and therefore must produce them themselves.

- Close networks of manufacturers and suppliers will be formed. It will become of decisive importance to bind the right partners to oneself early enough. The common goals must be clear, as must the allocation of roles, in order to create a new quality of cooperation. Automobile manufacturers will have to identify and expand promising business models in the network early on, to obtain competitive advantages. In addition, fields of strategic competence will have to be specifically strengthened and border areas will have to be integrated into trend-setting cooperations.

- In future the suppliers will bear the brunt of investments. But their capital base is already insufficient. Going public, investor groups or knock-on assistance from the automobile manufacturers are possible solutions. Suppliers will have to begin to work out financing strategies in good time.

- The relationship between manufacturer and supplier will become increasingly aggravated because the automobile companies will step up their attempts to pass on extra costs to their suppliers. However, because of the high costs, the suppliers will demand a contribution from the manufacturers. Until now it was above all the manufacturers with strong brands and powerful market positions who had an advantage over the suppliers on the price front. But due to weak consumption price power has declined here, too.

- The manufacturers´ component plants are becoming direct competitors of the suppliers. They work both for their group's own brands and for outsiders to the group. In the long-term they will only survive as component plants of the car manufacturers if they become providers of strategically important company competence and are of great importance for brand leadership.

4 Further concentration of automobile manufacturers by 2015

4.1 Predatory competition shows its effect

The history of the automotive industry has for decades been globally characterized by a steady decrease in the number of independent automobile manufacturers (see Fig. 60). While the reasons for the shrinking number of market players in the last century were at first purely of a microeconomic nature – product weakness, capital bottlenecks, management mistakes, etc. – and thus came from the companies themselves, and not from the macroeconomic market environment, the picture has changed very much in the last thirty years. With the advance of first the Japanese and then the Korean automotive industry in the global market at the same time as increasing saturation in the large volume markets of the triad a destructive competition began, of a kind not known before. This has continued to gain in fierceness until now – and will still continue to do so.

This crowding-out competition led to a process of concentration which with a shrinking number of market players (2004: 12 independent groups[53]) has meanwhile slowed down in textbook manner, as the remaining groups have become more similar in strength, and thus their competitive power and survivability have increased noticeably. Nevertheless, the process is not yet finished as long as that state of peace which characterizes the close oligopoly has not been reached.[54]

The IWK estimates that there will be 9 remaining independent groups in 2015 (Fig. 60).

---

[53] These are: BMW, DaimlerChrysler, Ford, GM, Honda, Hyundai/Kia, PSA, Renault/Nissan, Fiat, Toyota, VW

[54] Conf. FAST-2015 (2004).

Fig. 60. Concentration Automobile manufacturers 1960-2015

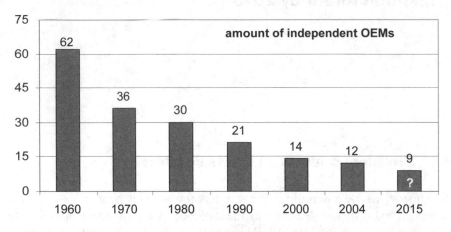

sources: automotive industry, IWK estimation

Before this background the strategic evaluation of the sustainability of the remaining OEMs as employers, investment targets, tax payers, etc. is of great importance to a large number of interested groups, such as banks, workforce, capital investors, suppliers, economy politicians, etc.

Unfortunately the significance does not correlate with a low degree of difficulty in establishing the sustainability of the companies concerned. Enterprise evaluations or enterprise ratings as such are a science of their own, which can only be performed professionally by a few specialized firms such as S&P, Moody's, etc, and are very costly in terms of both personnel and expenses. In order to carry out enterprise ratings the investigation, processing and evaluation of a large amount of objective information ("hard facts") and subjective assessments ("soft facts") are necessary, which then, in the overall view, make it possible to come to a judgement on this sustainability / survivability of the individual manufacturers or automobile companies. And to make matters worse, among the big automobile manufacturers there is none with one *single brand* – apart from Honda and Porsche. All the rest have two or more (BMW, Toyota, Renault, PSA) or a multitude of brands (GM, Ford, Fiat, Volkswagen), which are not only all more or less differentiated by regions or sectors (by market segments, lower, middle and upper), but above all are also variably successful within their groups (e.g. Audi in the Volkswagen group). [55] Today experts count 250-300 brands worldwide, of which most belong to the 12 biggest groups.

Making such extremely unequal automobile groups comparable is not possible without very restrictive assumptions and simplifications. Add to that the necessity to write the individual rating judgements in a way that allows a cardinal ranking of the participants. Here, too, are a great number of statistical pitfalls and a great many heroic assumptions must be made, making in the end any ranking of the companies vulnerable. This applies even if such an undertaking was carried through with the expense of the big rating agencies, which would have surpassed by far the possibilities open to the IWK.

The author has been aware of that from the start. Nonetheless, we believe that long-standing competence in the field, solid economic and business-administrative training and the grace of an awareness of one's own inadequacy are enough to arrive at valid results. In order to keep criticism of the calculations and method to a minimum, suppositions and methods are laid open, existing findings of the rating agencies evaluated, complemented by many quantitative company-specific data, external evaluations of the companies by specialists in the market, and our own qualitative strategic assessment of the future developments in the groups, based on experience.

All this information is weighed and evaluated cardinally and summed up in the **"IWK-Survival-Index" (ISI)**, which was specifically developed for this purpose. The order should – proceeding from the current strategic economic position of the companies – give information about the future market strength of the OEMs.

The "IWK-Survival-Index" (ISI) was developed in order to be better able to assess *competition resistance* and *survivability* of the automobile companies involved. The *ISI* is based on a consistent selection of company code numbers which are based on economic theories and empirically measurable branch experience. It signals how strongly an OEM is exposed to high competitive pressure and represents an indicator for the probability of whether a company has enough strength in the long-term to maintain an independent position in the market or will be taken over and eventually

---

[55]As far as the automobile line of business is only part of a combined group, ie. as at Fiat, the automobile line of business is treated as an independent enterprise with self-contained profit maximization.

disappears from the market. Remember: it is all about the companies as such, not about their individual brands, such as, for example, Audi in the VW group, or Volvo in the Ford group, etc. Even if a group disappears as an independent entity from the market, the brands have in the past usually survived under new ownership. Rover is inasmuch no exception to this, as this company was already gutted under the leadership of BMW, and as a company it only consisted of the Rover brand.

To prevent further possible misunderstandings: The ISI is a instant photo of today, not a value judgement for all eternity, that is the present ranking of the 11 manufacturers examined mirrors their future market strength, on the basis of knowledge available in the winter of 2004/2005. The example of the Chrysler group, which once already found itself standing at the edge of an abyss in the 80's and was then revitalized by Lee Iacocca in an unprecedented show of strength, shows that enterprise success is achieved by people with leadership strength, vision and charisma, i.e. by entrepreneurs, not by managers with semi-skilled management principles (management by...).

Enterprises which today show a low ranking position may arrive at a position of new market strength and improved survivability within one or two years, if fortune is guided by

- *the* ingenious and visionary enterprise manager with charisma and long-term visions

- and with the absolute support of shareholders interested in the long-term, rather than quarterly success

German automobile companies such as BMW offer positive examples of this in the post-war era. However, European and German automobile history can also offer the opposite examples of how flourishing enterprises are brought to the brink of ruin.

But the probability that this miracle of revitalization happens is not particularly great (see premises).

It is in any case true and also important that the "IWK-Survival-Index" is not static, but must be newly determined annually, if the latest evaluation status about the manufacturers is to be taken into account in the decisions of the interested groups – suppliers, investors, economic policy, etc.

Above all for suppliers, the ISI provides assistance – albeit only rough - in answering the following central questions:

- Which automobile group at the present date has the biggest chances of existing in the long-term in hard predatory competition?

- Which automobile groups do I have to gain as customers in order to be able to plan earnings development with as much security and as little disturbance as possible?

4.2 Evaluation of the 11 largest OEMs according to "survival-criteria"

4.2.1 Analysis method

The *survival capacity* of the individual automobile groups is determined by means of:

- code numbers describing the *current* economic situation of the group (CES),

- factors influencing the *future* development of the enterprise (=survivability)

Both these categories may be divided further into diverse sub-groups.

The current economic situation (CES) of a group is described by code numbers which provide information about the size of the enterprise, its current growth, as well as stability in the broader sense. For this the following information categories are evaluated:

> - market shares
> - growth dynamics
> - commercial value of equity capital
> - creditworthiness
> - productivity
> - profitability and
> - stability.

Forward looking statistical parameters show the strengths and weaknesses of a group with regard to future safeguarding, competitiveness and strategy of the enterprise. For this the following categories are evaluated:

- future orientation (investment activity),
- innovation orientation,
- effectiveness of R&D expenses,
- degree of globalization,
- utilization of potentials, image,
- effectiveness/ quality of management.

A further important aspect of the ranking is the subjective evaluation of enterprise strategy by the IWK. In this area all those influence factors which are not, or not usually) measurable quantitatively ("soft facts") are taken into consideration. In this the author's (leader of the IWK) experience of more than 30 years plays a decisive role.

Only those automobile companies were evaluated which are globally active or co-operate in the global market and showed a relevant market size on the basis of their sales volume and the number of units produced. [56]

Although a close cooperation and mutual participations exist between Renault and Nissan, the companies were considered separately in the IWK ranking, because their company reports are also still issued separately. In the same way, the analysis of Hyundai took place without consideration of Kia, because here, too, separate balances are issued.

The DaimlerChrysler group was analysed without Mitsubishi from the beginning, rightly, as has become apparent in the meantime. Further automobile manufacturers such as Mitsubishi, Mazda or Porsche were not considered due to their insufficient size.

For a uniform valuation of all automobile groups all reported enterprise code numbers are converted from Euros to US dollars at the following exchange rates (status: the end of the respective period):

---

[56] Altogether in 2004 acc. to Manager Magazin, world-wide there were 38 independent automobile manufacturers (edition 07/04, Champions League der Konzerne). Conf. the list in appendix 2.

Table 20. Exchange rates US$/Euro, end of the period

1995	1996	1997	1998	1999	2000	2001	2002	2003	2004
1.314	1.251	1.103	1.170	1.005	0.930	0.881	1.049	1.263	1.330

source: FERI

For information only

4.2.2   Establishment of the "IWK`s survival-index"

The construction of the ISI follows – in short – the following steps:

1. scale of indicators:

-   scaling of indicators on a point scale
-   indicator specific calculation of extreme or average value
-   allocation of indicator values on the point scale 1 to 100
-   allocation of the groups to point values according to index number

2. weighted summation by means of the index numbers of the branch:

-   weights correspond to the significance of the respective index number.
-   result: "IWK survival index"

4.2.3   Sources of information

The following information sources were used in the preparation of the *IWK survival index"*:

-   annual reports and companies' own studies

-   rating agency publications, e. g. Moody's

-   data bases of capital market portals, e. g. MSN, Comdirect, OnVista, Wallstreet-online etc.

-   ADAC; PWC, WestLB; VDA; Power

-   press and internet research; professional journals (e. g. Automobile Week, Automotive industry, Automobile Production, economy magazines, daily press etc.)

4.2.4 Valuation model

The following diagram is a schematic presentation of the "IWK survival index"

Fig. 61. Model diagram of the calculation of the "IWK survival index"

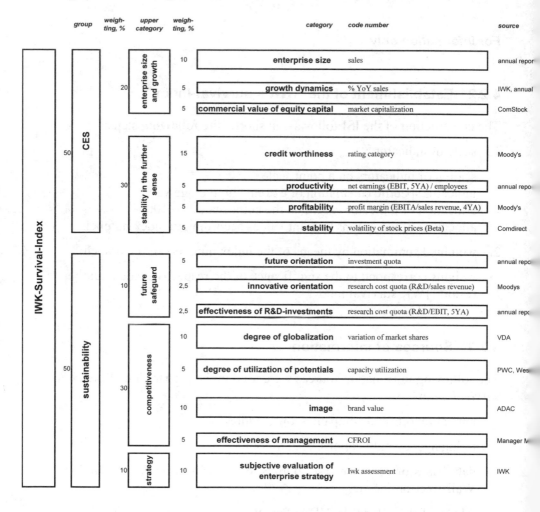

4.3 "IWK Survival-Index": valuations in the various categories

The following information is evaluated in establishing the *"IWK survival index"*:

4.3.1 Current Economic Situation (CES)

4.3.1.1 Enterprise size and growth

Size of Enterprise

For the evaluation of enterprise size the current turnover figures (in millions of US dollars) were consulted. This analysis of size is decisive for the evaluation of the survival chances of a company because it may be assumed that when a certain size of the enterprise is attained, independence cannot simply be given up, due to the economic and socio-political involvement of the enterprise. Thus the saying, "big is beautiful" is certainly valid here.

Growth dynamics

In order to be able to take into account the development of the enterprise over time the companies are evaluated on the basis of their growth dynamics. These were determined as the average percentage change in turnover in the past 5 years. The company with the highest growth value was given the maximum value of 100 points. Correspondingly, the company with the slowest growth, or that with the largest decline in turnover, was given zero points.

Commercial value of equity capital

Not only was the demand aspect on the commodities market drawn on in evaluating the current situation of an enterprise, but also the demand on the capital market. For a comprehensive assessment of size of an enterprise a code is necessary, which expresses the current market value of the company. The size of market capitalization is most suitable for this, as it mirrors the current stock market value of a listed company. The higher the market capitalization of an enterprise, the higher the interest of investors, as a rule. The result is that the groups with the lowest stock market value are the first in line for possible takeover by a bigger and financially better positioned group – if such a takeover fits in with strategic calculations. The high expenses incurred by BMW or General Motors to end a participation (Rover) or to avoid a takeover (Fiat) show clearly that this criterion only makes sense in the overall context.

Results in the category enterprise size and growth

The results in this main category are summarized in the following table:

Table 21. Results in the category enterprise size and growth

upper category	enterprise size and growth			total evaluation of the category	
category	enterprise size		growth dynamics		commercial value of equity capital			
code number	sales	Score	% YoY sales	Score	market capitalization	Score	total score	rank
unit	mil. US $		% YoY (4YA)		US $			
weighting factor	0,1		0,05		0,05			
source	annual reports	IWK	annual reports	IWK	ComStock	IWK	IWK	IWK
1. Toyota Motor Corporation	163.637,0	88	9,22%	64	129.418	100	85	
2. Honda Motor Co., Ltd.	78.222,0	42	8,69%	61	45.604	35	45	
3. BMW AG	52.446,1	28	4,92%	40	29.550	23	30	1
4. Nissan Motor Co., Ltd.	70.087,0	38	6,82%	51	45.139	35	40	
5. General Motors Corporation	185.524,0	100	2,75%	28	19.271	15	61	
6. Hyundai Motor Company	20.922,6	11	15,55%	100	13.252	10	33	
7. Ford Motor Company	164.196,0	89	0,30%	14	18.790	15	51	
8. DaimlerChrysler AG	172.319,9	93	-2,13%	0	41.262	32	54	
9. Peugeot S.A.	68.716,0	37	9,78%	67	14.142	11	38	
10. Volkswagen AG	110.074,2	59	4,02%	35	18.340	14	42	
11. Renault S.A.	47.394,1	26	0,15%	13	24.451	19	21	
12. Fiat S.p.A.	61.061,0	33	0,61%	16	6.917	5	22	
The best value	185.524,0		15,55%		129.418		85	
The worst value	20.922,6		-2,13%		6.917		21	

4.3.1.2 Stability in the further sense

The stability of an automobile company in the further sense is about the assessment of its liability to cyclical and external disturbances, for example crises on financial markets, or abrupt changes in the key dollar exchange rate. Does the group have sufficient reserves and above all the management skills to protect itself from external crises over a longer period of time, without losing its independence?

Creditworthiness

In judging the financial stability of an enterprise the evaluation of its creditworthiness is a decisive criterion. International rating agencies such as Standard & Poor's (S&P), Moody's and Fitch regularly check the creditworthiness of debtors, which inevitably undergoes constant change due to changes in the macroeconomic and enterprise specific environments.

In our analysis we have used as a base Moody's rating grade as indicator for the creditworthiness of an automobile company. This agency focuses its rating procedure on the determination of the earning power of an enterprise, that is, the ability to achieve future income surpluses.

The following aspects are taken into consideration by the analysis: [57]

- On the macroeconomic level one considers branch trends, like costs and prices, national and international competition as well as technological change in the automotive industry. These factors hold important information about the development of earning power, assets, financial needs and possible liabilities in the future. The political and regulatory environment in the country or countries where the company is active is also taken into account. From that the effects can be determined which arise from the general political conditions, the intensity of legal regulations and from money, tax and exchange rate policy. All these have consequences for the ability of an enterprise to meet its future obligations.

- At the level of the individual enterprise the quality of the management, though difficult to quantify, is one of the most important factors for the creditworthiness of an issuing company. Following aspects are analysed here: can the enterprise survive in competition; to what extent is financing scope exploited; what is the relationship to subsidiary companies, what is the relationship to regulatory authorities and what is its attitude to all the other factors which effect creditworthiness? Particular atten-

---

[57] Conf. Moody's: special report rating methodology for industrial enterprises

tion is paid to the comparison of entrepreneurial performance in the past with the management's future plans.

- Future earning power is additionally determined by a series of other factors. Firstly by the market positioning of the enterprise in comparison to its competitors. Especially the ability to achieve sales targets and at the same time keep costs under control is an important indicator. Secondly, the financial situation and the possibilities of finance and liquidity procurement play an important role. The same is true of the third point, group structure, whereby the emphasis is on the analysis of the various demands on the cash-flow of individual legal units within the group. Fourthly, support mechanisms such as guarantees and comfort letters of the parent company are relevant. And the fifth point is that the event risk must be assessed, that is, the danger that the fundamental creditworthiness of a listed company could worsen due to mergers, acquisitions or leveraged buy-outs, or because of judicial, legal or regulatory measures, or even due to an industrial accident.

The current valuations of OEMs by the rating agency Moody's as of September 2004 are presented in Table 22.

Productivity

In this category economic efficiency, output capacity and competitiveness of a company are determined. An appropriate index figure for this is productivity, calculated as the relation of earnings before interest and taxes (EBIT) per employee. This figure gives information about the efficiency of the factor labour, or the profitability of the employed personnel. The higher the earnings per employee the more productively the company is and the more chances it consequently has of surviving in the competition. Since the EBIT is subject to strong fluctuations over time – depending on the company's balance policy – the IWK Index uses an average value from the last 5 years in the calculation of this code number.

Profitability

An index figure which gives information about profitability is the percentage return on sales, or the so-called EBITDA margin (EBITDA = earnings before interest, taxes, depreciation, and amortization). This figure and its relation to turnover is very suitable as a relative index figure for the comparison of the earnings power of different companies (especially internationally). It shows the percentage of operative profit before depreciation a company was able to produce in relation to sales volume. Since the figure

EBITDA which is drawn on is also subject to strong fluctuations over time, an average of the last 4 years is used here, too.

Stability in the strictest sense

The volatility of share prices gives information about the stability and "robustness" of the enterprise with regard to the short-term market conditions. The assumption is that the capital market and stock exchange react fastest to specific company events, which is then mirrored in the fluctuations of the share prices. If a company is developing stably (foresee ably), these movements will tend to be small, and comparable to the market as a whole. But if the stock exchange does not seem to show any great confidence in the company and if the smallest event can make a significant change to the market value of the company, that is as a rule a sign that the company is relatively unstable.

For the IWK analysis the stability of the OEMs' share prices is evaluated using two index figures. On the one hand is the volatility, which shows a statistical measurement of risk and measures the fluctuation intensity of the price of an underlying instrument within a certain period. The higher the volatility, the more strongly the price swings up and down, and the more risky investment in the underlying instrument is. In order to relate the stability of the individual share price to the general market development the IWK analysis also applies the *beta factor* of the share price as a second index figure. This is a statistical measure which represents the relative range of movement of a share price in relation to the market as a whole. It measures the sensitivity of a security with regard to the relevant market index (e.g. DAX, Nikkei or Dow Jones). A security with a beta factor of 1 thus shows a level of fluctuation comparable to that of the market index, the nearer the beta factor is to zero, the smaller the fluctuation in comparison to the market as a whole.

Both these index figures, volatility and beta factor, were determined for the last 3 years for the share prices of the companies studied and were fed into the IWK evaluation with the same weighting. For better understanding let it be expressly mentioned that in this category the stability of share prices is judged, and not successful performance. As far as the definition goes, a large rise in share price (as Hyundai: more than 100% in the past three years) means high volatility and is therefore a negative point in the evaluation, since this mirrors (albeit positive) instability and the positive development is already accounted for in other categories.

Results in the category stability in the broadest sense

The results in this main category are summarized in the following table:

Table 22. Results in the category stability in the broadest sense

code number	credit worthiness rating category (rating category, 0.15, Moody's)	score (IWK)	stability in further sense — productivity: net earnings (EBIT, 5YA)/employees (US $, 0.05, annual reports)	score (IWK)	stability in further sense — profit margin (EBITA/sales revenue, 4YA) (%, 0.05, Moody's)	score (IWK)	stability — volatility of stock prices (Beta) (Beta, 0.05, Comdirect)	score (IWK)	total evaluation of the category — total score (IWK)	rank (IWK)
1. Toyota Motor Corporation	Aaa P-1 STA	100	36,555	91	8,4%	99	96,92	100	98,2	1.
2. Honda Motor Co., Ltd.	A1 P-1 STA	77	37,597	93	7,9%	93	83,34	86	83,9	2.
3. BMW AG	A1 P-1 STA	77	34,974	87	6,9%	81	48,54	50	74,8	4.
4. Nissan Motor Co., Ltd.	Baa1 P-2 STA	62	33,493	83	8,5%	100	82,47	85	75,7	3.
5. General Motors Corporation	Baa1 P-2 NEG	60	40,330	100	1,5%	18	46,53	48	57,6	6.
6. Hyundai Motor Company	Ba1 REV UP	49	36,081	89	7,1%	84	22,56	23	57,2	7.
7. Ford Motor Company	Baa1 P-2 NEG	60	34,410	85	1,9%	22	13,08	13	50,2	10.
8. DaimlerChrysler AG	A3 P-2 STA	67	14,824	37	2,4%	28	29,71	31	49,4	11.
9. Peugeot S.A.	A3 P-2 POS	69	12,859	32	5,0%	59	65,86	68	60,9	5.
10. Volkswagen AG	A3 P-2 NEG	65	13,741	34	3,4%	40	36,59	38	51,1	9.
11. Renault S.A.	Baa2 P-2 REV UP	59	18,836	47	4,5%	53	53,93	56	55,4	8.
12. Fiat S.p.A.	Ba3 NEG	35	5,402	13	0,0%	0	47,60	49	27,9	12.
The best value		100	40,330		8,5%		13,08		98,2	
The worst value		35	5,402		0,0%		96,92		27,9	

4.3.2 Sustainability

4.3.2.1 Safeguarding the future

To safeguard a successful future for an automobile company

- investment quota,
- innovation orientation and
- expenses for research and development

play a decisive role. All these factors have an impact on the enterprise rating.

Future orientation

The future orientation of an enterprise may be judged best, according to general opinion, by means of its investment policy and investment tendency. Both are reflected in the investment quota. This is calculated as the percentage share of investments in plant and equipment, in relation to fixed capital. Since the level of investment is usually an expression of growth oriented company strategy, a high or increasing capital outlay ratio reflects not only a positive business development in recent years, but also and above all positive business expectations for the future. In order to avoid cyclical and balance fluctuations in the capital outlay ratios and thus render the OEMs more mutually comparable the analysis uses an average of the respective capital outlay ratios of the last four years as a basis.

It may be conceded to critics that the capital outlay ratio is a good indicator for the growth orientation of an enterprise, but that in the end it does not permit any conclusion about quality, efficiency or cost-effectiveness of an investment. Vacant automobile factories or those which can only be run at a loss confirm this perception.

Innovative orientation

It is not new knowledge that innovative companies have a competitional advantage. But especially in the highly technologized automotive industry with its current process of crowding-out and concentration leadership in the field of innovations is of existential importance as a marketing instrument for the OEMs. A company can only secure a winning technological margin and thus create the possibility of withdrawing from the price competition at least temporarily by constant innovations. Therefore research and development are of decisive importance in the automotive industry. And it is also why BMW only managed to become competitive on the global market in the 80' with the creation of an independent research and engineering center (FIZ).

Research and development (R&D) are defined as follows:

- *Research* is the generating of new scientific-technological knowledge and its combination with already existing knowledge to gain new knowledge, which at least in the long-term may serve the enterprise as a basis for innovations.

- *Development* is the transformation of the demands of the market in connection with new scientific-technological knowledge, gained from research, into marketable products and processes.

How well positioned an enterprise is with regard to the innovation challenge can be judged on the basis of its R&D expenditure. If this figure is set in relation to revenues information about the innovative orientation of a company. The research cost quota thus shows what proportion of the revenue the company re-invests in research and development, how seriously it takes its sustainability.

However this indicator must be handled with care inasmuch as the largest part of innovations in the automotive industry today is produced by the components supply industry and as the case may be, co-financed by OEMs.

Effectiveness of R&D investments

A significant disadvantage of the *research cost quota* is that no conclusions about the purposefulness and effectiveness of R&D investments can be made. For the OEM a reasonable balance between cost-efficiency and innovation leadership is essential. In practice there are numerous examples of excessive investments having led to the wasting of company resources. Only those companies which have their costs under control, including and especially the R&D costs, are durable and sustainable.

The effectiveness of R&D investments is very difficult to quantify. Sufficient insight is gained here if the R&D expenditure is set in relation to the earnings before interest and taxes (EBIT) of the last 5 years. This research cost quota shows which proportion of trading results the company re-invests in research and development. This in turn allows – very careful - conclusions about research efficiency.

Results in the category future safeguarding

The results in this main category are summarized in the following table:

Table 23. Results in the category future safeguarding

upper category	future safeguard						total evaluation of the category	
category	future orientation		innovative orientation		effectiveness of R&D-investments			
code number	investment quota	score	research cost quota (R&D/sales revenue)	score	research cost quota (R&D/EBIT, 5YA)	score	total score	rank
unit	%		%		mil. US $			
weighting factor	0,05		0,025		0,025			
source	annual reports	IWK	Moody's	IWK	annual reports	IWK	IWK	IWK
1. Toyota Motor Corporation	5,07	99	4,3	67	74,0%	85	88	4.
2. Honda Motor Co., Ltd.	4,49	95	5,5	86	90,8%	77	88	3.
3. BMW AG	4,18	92	6,4	100	92,0%	76	90	1.
4. Nissan Motor Co., Ltd.	5,22	100	5,0	78	84,5%	80	90	2.
5. General Motors Corporation	2,07	77	3,7	58	52,4%	97	77	7.
6. Hyundai Motor Company	3,79	89	4,1	64	46,2%	100	86	5.
7. Ford Motor Company	2,43	79	5,4	84	76,9%	84	82	6.
8. DaimlerChrysler AG	-4,24	30	4,6	72	147,7%	47	45	11.
9. Peugeot S.A.	2,06	77	4,0	63	106,9%	68	71	9.
10. Volkswagen AG	2,06	77	5,4	84	133,6%	54	73	8.
11. Renault S.A.	-0,80	55	4,9	77	96,7%	74	65	10.
12. Fiat S.p.A.	-8,23	0	4,0	63	237,4%	0	16	12.
The best value	5,22		6,4		46,2%		90,2	
The worst value	-8,23		3,7		237,4%		15,6	

4.3.2.2 Competitiveness

Degree of globalization

Globalization has become the key word in many areas in recent years, and as we have seen, the automotive industry was among the pioneers. In the modern world only those companies are successful which are present on all markets in a shrinking world. The dispersion of cyclical sales risks and the diversification of market and currency risks play a big role here. An evenly spread presence in the world's most important markets is an indicator of competitiveness and a balancing of cyclical, market and exchange rate risks and thus of sustainability.

In order to determine the worldwide diversification of an OEM for the purposes of the *ISI*, the current passenger car market shares of the OEMs in the triad are analysed. The criterion is a best-possible balancing of the size of market share in the triad, that is, that a manufacturer is evenly represented in all the important markets and is not excessively dependent on any individual region. The greater the deviation is from such an evenly spread market presence, the worse the valuation of the degree of globalization of the company will be. Econometrically this index figure is measured as a co-efficient of variation of the market shares in the USA, Western Europe and Japan.

The coefficient of variation is calculated from the standard deviation of a spot check, divided by the arithmetical mean of this spot check.

$$V = \frac{s}{x} \text{, whereas } s = \sqrt{s^2} = \sqrt{\frac{1}{n}\sum_{i=1}^{n}(x_i - \bar{x})^2}$$

Thus the coefficient of variation makes the comparability of dispersion values with varying arithmetical averages possible by means of the relation of the individual standard deviations to their individual arithmetical means.

Rate of capacity utilization of manufacturing plants

Automobile production (above all pressing plants, bodyshell work, paint shops etc.) is very capital intensive and requires high investments. To be able to fulfil the prerequisite of a successful automobile company - profitability – capacity must be utilized as highly and as constantly as possible. A measurement for this is the degree of capacity utilization. Demand cycles and internal model cycles lead to strong fluctuations in the capacity utilization of the companies. In the IWK evaluation the degree of capacity utilization of the OEMs, measured at a fixed time (i.e. in the same cyclical phase), was used as the index figure for competitiveness, since companies with a higher degree of utilization show a lower susceptibility to economic swings and thus better competitiveness.

Image

The brand value of an OEM can essentially be recognised from the factors customer satisfaction and image of the company in question. Appropriate surveys for the German-speaking regions are carried out, for example, by ADAC. In order to evaluate the international comparability of the OEMs the results of the market research company J.D. Power, which are recognized throughout the branch, were drawn on for the IWK evaluation.

The customer satisfaction studies consist essentially of the components "quality and reliability", "ownership costs", "service satisfaction", and "vehicle appeal". For the Western European region an average value of the results for Germany, Great Britain and France was determined, and additionally the results for the USA, as the largest market, were included with the same weighting in the IWK analysis. For companies which are represented with several brands in the respective countries, a weighted average of the individual company brands was determined. Since the companies Renault, PSA and Fiat are hardly represented in the US market and thus no estimates for customer satisfaction there exist, the average values for the branch as a whole were used for these three groups.

Brand value in the form of customer satisfaction and image is an important measurement for future competitiveness, as it has a decisive influence on customer retention and the brand loyalty of car buyers. In addition, high customer satisfaction possesses an important "radiation" effect for potential new customers, and it is not without reason that manufacturers like to use good results in the corresponding surveys as a persuasive instrument in advertising.

Effectiveness of the management

The assessment of the effectiveness of enterprise management plays an important role in judging sustainability. This is usually valued qualitatively, but an indirect quantification is possible to a certain extent on the basis of financial management figures, and is undertaken here.

The effectiveness of management is expressed as CFROI (cash flow return on investment) and relates to the latest published business figures. The CFROI allows conclusions to be made about how effectively the management has employed the working capital, that is, how high the liquidity returns generated by this capital are.

The *CFROI* is more significant than the *EBIT* because:

- it relates exclusively to sales revenue and therefore is less susceptible to the accounting methods, which distort the EBIT;

- it takes into consideration the assets which generate sales revenue;

- it considers the non-balanced assets (so that enterprises, which lease their sites don't have an advantage over to those which are the owners of their sites);

- it calculates the useful lifetime of an asset and a mixture of lapsing values and non-lapsing values;

- it is inflation-adjusted and thus not nominally distorted.

Results in the category competitiveness

The results in this main category are summarized in the following table:

Table 24. Results in the category competitiveness

upper category	competitiveness								total evaluation of the category	
category	degree of globalization / diversification		degree of utilization of potentials		image		effectiveness of management			
code number	distribution of market shares	score	capacity utilization	score	brand value	score	CFROI	score	total score	rank
unit	coeff. of variation		%		mark		%			
weighting factor	0,1		0,05		0,1		0,05			
source	VDA	IWK	PWC, WestLB	IWK	ADAC	IWK	Manager Magazin	IWK	IWK	IWK
1. Toyota Motor Corporation	0,823	60	85	89	79,8	99	7,5	58	77	3.
2. Honda Motor Co., Ltd.	0,690	73	85	89	80,9	100	8,9	66	84	1.
3. BMW AG	0,604	82	96	100	76,7	95	5,4	45	83	2.
4. Nissan Motor Co., Ltd.	0,649	77	74	77	44,3	55	14,5	100	74	4.
5. General Motors Corporation	0,428	100	74	77	45,7	57	5,5	45	73	5.
6. Hyundai Motor Company	0,736	68	71	74	48,0	59	9,2	68	66	6.
7. Ford Motor Company	0,762	66	71	74	52,2	65	2,8	29	60	7.
8. DaimlerChrysler AG	0,730	69	75	79	46,6	58	1,4	20	59	8.
9. Peugeot S.A.	1,361	4	82	85	43,2	53	5,0	42	41	10.
10. Volkswagen AG	1,041	37	76	79	43,2	53	2,1	25	47	9.
11. Renault S.A.	1,404	0	74	77	47,1	58	1,2	19	35	11.
12. Fiat S.p.A.	1,361	4	65	67	32,6	40	-2,0	0	26	12.
The best value	0,428		96		80,9		14,5		84	
The worst value	1,404		65		32,6		-2,0		26	

4.3.3 Strategy

In this category all those influence factors are evaluated, which are not measurable by quantity, or only under certain conditions. The reflection behind this is that not only the actually determined (thus historical) reference numbers such as ,for example, those from balance and profit and loss accounts (PaL), must be quoted, but also others, which are more significant characteristics (qualitative risk and chance factors). These include above all:

- orientation towards the future and sustainability of strategic decisions taken on the development of the group, as they appear in the media and specialist press,

- sustainability of group policy,

- management structures,

- flexibility when faced with market changes,

- systematic quality assurance,

- marketing structures,

- information policy, leadership ethics, environmental awareness, etc.

In many cases it is just these qualitative factors which act as an early warning system in signalizing critical developments or sources of danger with regard to the sustainability of a company. For the purposes of our analysis such factors are thus very significant. However, their consideration is only feasible with criteria which are subjective and difficult to norm. A quantitative presentation is often impossible.

In order to secure the highest possible objectivity and not to unduly influence the results, such influence factors arc used with a weighting of only 10% in the IWK-Survival-Index. Sensitivity calculations by the IWK have shown that in bigger revaluations the final ranking of a company is only marginally influenced. The subjective evaluation of the companies is based on internal research, press reports, and consultations of experts, which were summed up to an overall view in the form of strengths-weaknesses-analyses for the individual OEMs.

Results in the category "Subjective evaluation of enterprise strategy" are presented in the following table:

Table 25. Results in the category subjective evaluation of enterprise strategy

	Strategy	
upper category		
category	subjective evaluation of enterprise strategy	
code number	IWK assessment	Score
unit	Score	
weighting factor	0,1	
source	IWK	IWK
1. Toyota Motor Corporation	100	100
2. Honda Motor Co., Ltd.	77	77
3. BMW AG	95	95
4. Nissan Motor Co., Ltd.	73	73
5. General Motors Corporation	50	50
6. Hyundai Motor Company	68	68
7. Ford Motor Company	55	55
8. DaimlerChrysler AG	62	62
9. Peugeot S.A.	75	75
10. Volkswagen AG	58	58
11. Renault S.A.	55	55
12. Fiat S.p.A.	40	40
The best value	100	
The worst value	40	

4.4 Total results of the ranking

The results of the individual evaluation categories and the IWK-Survival-Index for each OEM are summarized in the following table:

Table 26. Total survey. IWK survival index

| | | | CES | | | | | | | | sustainability | | | | | | | |
| | | | enterprise size and growth | | | stability in the further sense | | | | future safeguard | | | competetiveness | | | | strategy |
Company	RANK acc. to IWK-Survival-Index	total score	enterprise size	growth dynamics	commercial value of equity capital	credit worthiness	productivity	profitability	stability	future orientation	innovative orientation	effectiveness of R&D-investments	degree of globalization / diversification	degree of utilization of potentials	image	effectiveness of management	subjective evaluation of enterprise strategy
Toyota Motor Corporation	1.	88,39	88	64	100	100	91	99	100	99	67	85	60	89	99	58	100
Honda Motor Co., Ltd.	2.	75,76	42	61	35	77	93	93	86	95	86	77	73	89	100	66	77
BMW AG	3.	71,84	28	40	23	77	87	81	50	92	100	76	82	100	95	45	95
Nissan Motor Co., Ltd.	4.	69,07	38	51	35	62	83	100	85	100	78	80	77	77	55	100	73
General Motors Corporation	5.	63,88	100	28	15	60	100	18	48	77	58	97	100	77	57	45	50
Hyundai Motor Company	6.	59,04	11	100	10	49	89	84	23	89	64	100	68	74	59	68	68
Ford Motor Company	7.	57,14	89	14	15	60	85	22	13	79	84	84	66	74	65	29	55
DaimlerChrysler AG	8.	53,99	93	0	32	67	37	28	31	30	72	47	69	79	58	20	62
Peugeot S.A.	9.	52,65	37	67	11	69	32	59	68	77	63	68	4	85	53	42	75
Volkswagen AG	10.	51,06	59	35	14	65	34	40	38	77	84	54	37	79	53	25	58
Renault S.A.	11.	43,40	26	13	19	59	47	53	56	55	77	74	0	77	58	19	55
Fiat S.p.A.	12.	26,12	33	16	5	35	13	0	49	0	63	0	4	67	40	0	40

4.5   Conclusion: who has the best chances in the predatory competition?

The ranking results of the 11 biggest automobile groups uncover the substantial differences between the OEMs concerning their survival capacity in global destructive competition.

In summary the following perceptions may be put on record:

- In the evaluation criterion *market growth* and *growth* it is striking that large enterprises tend to have smaller growth dynamics. The OEMs with the biggest sales figures, **General Motors, DaimlerChrysler** and **Ford** are among the companies with the smallest average growth rates in the last 4 years. The company with the smallest sales figures, **Hyundai**, is in contrast, the strongest growing company by far. **Toyota** is an exception among the big companies, as are the other two Japanese OEMs, **Honda** and **Nissan.** They show above average growth dynamics. **Peugeot** is the only European manufacturer that can keep up with the others in this field.

  Significant knowledge can be gained from the category *shareholder value*: the market value of **Toyota**, the top company, is three times as high as that of second-place Honda and Nissan. The three Japanese companies therefore show a much higher market capitalization than all 9 other manufacturers together (!).

- In the category *stability in the broadest sense* the enterprises were evaluated concerning their credit worthiness, productivity, profitability and stability. Credit worthiness was given the biggest weight. The results in this category correspond to the final result of the IWK-Survival-Index for the top 3 companies. The **American companies** come off distinctly worse in this category.

  This result is confirmed in the category *profitability,* too: while the **Japanese groups** show the highest profit margins, these values – apart from **Fiat**, whose sales profit was near to zero in recent years – are worst for the US companies.

  Where productivity is concerned the enterprises can be divided into two groups: the highly productive companies, **Hyundai, General Motors, Honda, Toyota, BMW, Ford, and Nissan**, make an average

EBIT of 30 to 40 thousand Euros per employee. These values are well under 20 thousand Euros for the other manufacturers.

If one evaluates the stability of the groups on the basis of volatility and the beta factor of the share prices, it is again the **Japanese OEMs** which come out clearly on top. This testifies to a high level of continuity, consistency and predictability, management qualities which are honoured by investors.

- In evaluating sustainability, *investment activity* and *research and development expenses* and their *efficiency* are compared in the category future safeguarding.

  The investment activity of the groups shows clear geographical differences. The highest quotas are to be found in the **Japanese firms**, followed by **Hyundai**. Then come the **American groups** and finally the **European producers**. **BMW** is the only exception in this grouping, since it is only just behind the Japanese OEMs. **Renault, DaimlerChrysler and Fiat** even have negative investment quotas in the average for the last 4 years – not a good signal for future safeguarding.

  Investment in R&D is analysed on the basis of *absolute size* and *effectiveness*. The calculations have shown that there is no connection between size and efficiency that means that not the company which has invested most in research has also the highest pay-back on these investments. Research expenses as a proportion of sales revenue are very high in the companies **BMW, Honda, Volkswagen** and **Ford**. **GM Fiat** and **Peugeot** have the lowest research quotas. Altogether, the spread of investment quotas is very narrow, between 3.7% and 6.4%.

  As far as the *effectiveness of R&D investments* is concerned, **GM, Toyota,** and **BMW** show the highest values, in comparison to **Volkswagen, DaimlerChrysler** and **Fiat** at the other end of the scale. This result is not surprising, considering costly development policy on the one hand and various flops in the area of model policy on the other.

- One of the most significant criteria in the whole ranking is *competitiveness*. For this reason it is given an above average weight in the IWK study. the groups are evaluated in this category on the basis of their *degree of globalization, capacity utilization, brand value*, and *management skFig*. **General Motors, BMW, Nissan,** and **Honda** are the most regionally diversified, the **French and Italian producers** come off worst here, because they are so fixed on Europe.

With regard to *capacity utilization* **Toyota, BMW, Honda**, and **Peugeot** have above average values. The **Fiat, Ford** and **Hyundai** works were worst utilized at only 65% and 71%.

Looking at *market value*, **Honda, Toyota** and **BMW** show the best results in image and customer satisfaction. The other manufacturers follow at a distance, but are relatively close together. **Fiat** brings up the rear alone in this category.

Nissan has the *most effective management*, well ahead of the other manufacturers with a CFROI rate of 14.5%. The other three Asian groups, **Hyundai, Honda and Toyota** show rates in high single figures, however, and the European OEMs are at the bottom of the list. This mirrors on the one hand the difficult sales situation of these producers, which comes from the crowding-out competition described, and from management mistakes above all in the field of model policy. On the other hand it mirrors heavily loss-making acquisition and diversification attempts in the past.

- *Company strategy* was judged subjectively by the IWK. This is where all those influences were taken into account, which are not, or only to a limited degree, quantitatively measurable. **Toyota** takes first place, with its long-term company strategy geared to self-financing, its exemplary production system (TPS) and its strictly consumer-value oriented company philosophy. At present it is the measure of all things in the automotive industry. The **BMW group** follows in second place, which only landed second mainly because of its failed Rover and BRR acquisitions. **Fiat and GM** are again to be found at the bottom of the list because the company policy of both OEMs obviously suffers from being strategically very weak.

- The clear favourites in the overall evaluation, the **Japanese companies, Toyota, Honda and Nissan**, stand out geographically from the others. The winner of the ranking - **Toyota** – is at the front in almost all the evaluation categories and achieved altogether the most points by a long way. The **Korean manufacturer, Hyundai**, is on the best path to copy the success story of the Japanese OEMs. Based above all on the (still) small size of the company and the rapid but risky expansion, Hyundai already makes it to a good place in the midfield of the overall ranking.

- In the midfield are also the **American groups, GM and Ford**. These companies were able to convince mainly by means of their size and a high degree of globalization, but they show distinct weaknesses in profitability, management effectiveness and general company strategy. In addition, the American automobile giants show the lowest growth rates in the last few years.

- With the exception of **BMW** the **European groups** in the lower segment of the ranking. This is partly due to a current weakness in earnings, and partly to strategic failings in model policy and / or in the degree of globalization. Similarly to the *American producers*, **Daimler-Chrysler** only has its size and degree of globalization as trumps, followed closely by **PSA** and **Volkswagen.** Despite its small size, **PSA** profits mainly from its brilliant model strategy and strict management as a family enterprise. At the bottom of the overall evaluation come **Renault** and – far behind – **Fiat.**

- If one differentiates according to the purely German producers, it is remarkable that **BMW** takes third place, penetrating the group of Japanese firms, despite its relatively small size. BMW is above all convincing as to sustainability because of its good performance: in the categories innovation orientation and capacity utilization BMW is the absolute front-runner, in future orientation and brand value the company is in the top group and in the other evaluation criteria, too, BMW shows above average values.

   Thus **BMW** is the clear winner among the European automobile companies. Apart from considerable strategic weaknesses in the 90's, the company has had a spotless track-record since 1960 and the take over of entrepreneurial responsibility by the Quant family. Of all the European companies, BMW currently shows the greatest degree of survivability, well ahead of **DaimlerChrysler, PSA** and **Volkswagen.** For the sake of fairness, however, it must be added that it was mainly size which led to this placing for **PSA**, for whom brilliant strategic management as a family business in the style of Toyota or BMW opens up a distinctly better perspective for the future.

- Conversely, **DaimlerChrysler** is an example of how a previously front position as a premium producer can be wasted by continual mismanagement over years. Together with the other German company, VW, it is in the bottom half of the rating – with a very similar overall score. If one compares the German companies with the French companies, **PSA** and **Renault**, it becomes clear that the German companies have an advantage in their much higher degree of globalization and their larger size. Even though **Renault** only reached the second worst rating overall, when the company's participation in **Nissan** is taken into consideration its future prospects look better than its actual rating would suggest.

- At the bottom of the table, by a long way, is the **Fiat group**. In all categories apart from *company size, innovation orientation* and *stability* the Italian automobile manufacturer has the worst results. The Fiat group is probably the least equal to the challenges of the future crowding-out competition, despite its quite attractive brands. If it proves that there are no possibilities to revitalize the business after leaving General Motors, and if there is no chance of a change in its strategic orientation, the Italian firm will probably be the next to disappear from the market as an independent manufacturer. Whether the brands Alfa/Maserati and Lancia would then survive independently is more than questionable, given their small size and lack of global positioning.

5 Consequences for the Components Supply Industry

The relationship between manufacturer and supplier worsened clearly in 2004, because the automobile groups reinforced their attempts to pass on extra costs to the suppliers. However, the suppliers demand a contribution from the manufacturers because of the high cost burden. Previously it was above all the manufacturers, with big brands and a position of power over the suppliers, who had the advantages on the price front. Due to the lull in consumption however, price power has shrunk here, too. The question is how the long-term relationship between OEMs and suppliers will change in the future as a result of long-term development tendencies.

5.1 Typology of the added value chain

The upheavals which are to be expected at the manufacturer level as a result of global crowding-out competition will have varying effects on the suppliers, depending on their position and relationship structure with regard to their purchasers –OEMs or other components suppliers of a "higher order".

In order to better understand the functional chain it makes sense to typify the supply companies in the automotive industry by their position in the automobile value creation chain. Thus it is first necessary to present the current structure of the value creation chain and the various levels (1st tier, 2nd tier, 3rd tier, etc.) at which the companies are active. The pyramid form is generally known (Fig. 62).

Fig. 62. Supply pyramid

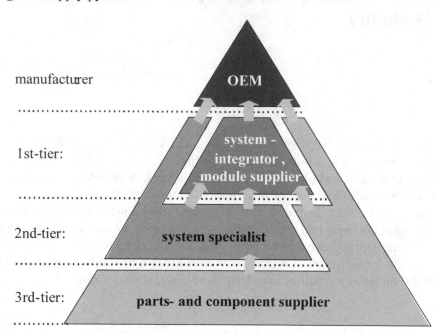

source: IWK presentation

Following the typology of the suppliers in Fig. 62, there now follows a description of the essential characteristics and tasks of the different levels of the value creation chain.

The *parts and components supplier (3ʳᵈ tier)* is characterized by both a relatively small installation activity and low development activity. As a rule it produces standard parts developed by the purchaser, according to exact specifications, and which can generally be produced without a particularly high level of technology.

Due to the hierarchical structure of the supply chain, the parts and components supplier's purchaser is increasingly seldom the OEM itself, as a rule the 3ʳᵈ tier supplier acts as a sub-supplier for another supplier. Then the individual products are taken by a superior module supplier or system specialist and put to their intended use within a module or system.

The system specialist (2ⁿᵈ tier) commands high technological development competence and low installation activity. As defined earlier, diverse assembly groups, aggregates and components are put together as a functional compound, but not necessarily as an installation unit. The special

mark of the system specialist is its ability to come up with creative technological solutions which in niche segments give it – at least temporarily – a
monopoly. These product innovations are initiated independently by the
system specialist, that is, not by order of the OEMs, but at its own risk.
The building and expansion of its own development potential requires a
readiness and ability to invest larger sums.

In comparison to the system specialist the module supplier (1st tier) is
generally less active in development and more extensively active in assembly. This means that it carries a lower development risk, but instead
takes on the entire logistics responsibility for the completion of the module
in question. A high level of production competence is the prerequisite for
this. The module supplier joins the parts and components from the subor-
dinate suppliers together to a module which is ready to be installed. This is
then delivered just in time (with a buffer) directly to the automobile manufacturer's assembly line.

The system integrator (1st tier) is at the very top of the supply hierarchy.
It commands a high level of production-related integration competence and
high technological competence. It takes on pre-assembly tasks as well as
significant parts of the development of pre-finished modules, that is, it
guarantees the whole system and thus also part of the market risk. So today
the system integrator is given tasks which were for many decades core
business and as such only performed by the manufacturer.

However, it must be mentioned that in practice individual suppliers can
be assigned to various positions at the same time within the supply pyramid, because of their broad product programs. They can fulfil the requirements of a system specialist with one product while appearing as a parts
supplier for another. Thus it is not always possible to designate a supplier
to one category or another. In the following therefore, the consequences of
the crowding-out competition at OEM level are described for the supply
industry as a whole and then differentiated according to the particular
range of tasks and the position of the company.

5.2   Concentration in the producers' oligopoly brings risks for suppliers

5.2.1   Upward pressure on costs from OEMs becomes still more rigid

The structure of the automotive industry is characterized by the pyramid form described above, in which OEMs are the uppermost and last level of the real added value chain. The intermediate products are delivered to them for final assembly from the suppliers, who are themselves supplied by companies at a lower level. This structure extends over several levels, where clear dependencies between the individual levels also exist.

On the other side of the market, opposite this entire added value chain (incl. trade) [58], there is the new vehicle purchaser as the final customer. Therefore it is unavoidable that stagnating or even shrinking sales of the final product will affect all the players on the supply side. The crowding-out competition and cost pressure among the automobile manufacturers mentioned above continues undiminished to the downstream levels since the whole power structure within the added value chain is changed. Therefore the whole process of the cost competition, including the necessary rise in quality and adaptation to customer wishes, applies in the same way to the suppliers. Their customers are in this case their direct purchasers, that is, the manufacturers or intermediate suppliers. This focusing on the final customer thus permeates the whole added value chain.

Due to market pressure the automobile manufacturers have for years been unable to pass on their costs to the final customer in the usual form of effective price rises, but are impelled to drastic cost management. They roll back a large proportion of their internal cost problems down to their suppliers, by passing on the margin pressure to which they themselves are exposed directly to their suppliers. The price pressure at the 1st tier level is passed on as a cascade to the downstream supply levels (cascade competition), which in the end must lead to withdrawal from the market (merger, bankruptcy) by borderline suppliers at every level.

The *product specialists, innovation pioneers* and *niche suppliers* are able to avoid this pressure from the OEMs the longest. But only for as long

---

[58] Trade by the following consideration because of simplicity will be neglected, although it perceives at first the tight market.

as the OEMs as purchasers value the benefit advantages of cost-intensive innovations – which help them to temporarily unique positions and thus to competition advantages in the fight for customer favour in the market – more than the additional price. However, the possibility of escaping the cost scourge of their purchasers, whether OEMs or system suppliers, by means of innovative products is becoming more and more difficult for the suppliers, due to the high capital expenses necessary for research and development. Sooner or later they also come under pressure; in the end no link of the added value chain can evade the necessity to reduce costs – the imitative competition from other suppliers in the world market takes care of that.

Considering the mega-trends in the global market which were described in chapter 3, the pressure on suppliers from the manufacturers will continue to increase. The cost screw must be tightened year by year to be able to sell products and services at more and more favourable prices. The price pressure from their purchasers thus leads to decreasing profits for the downstream suppliers, with the result that these are hardly able to secure their own sustainability any more by means of research and development for innovative products.

This development conflicts with the plans of the OEMs to increasingly turn their suppliers into development partners. Because of the increasing trend of outsourcing the production and development activities of the manufacturers a further structural transfer of costs from the manufacturers to the suppliers occurs. Many suppliers cannot afford the higher investment and organizational costs necessary and are taken over by more financially powerful competitors – also increasingly by financial investors.

Additionally, the automobile suppliers face a greater challenge than other firms to be close to their customers worldwide with their production sites, in order to be able to deliver to the assembly line just in time or just in sequence and at optimal cost. Supply firms which are in such direct contact with the manufacturing plants are therefore compelled to improve their supplier management and to align their procurement, production and logistics systems securely and stably. The capital-intensive building of production capacities and international logistics management is indispensable for 1st tier suppliers in times of OEM globalization, but demands high advance investments, for which the OEMs do not compensate them. This often overstretches both financial resources and organizational structure of the suppliers, which as a rule are medium-sized firms.

The consequences are discontinuance of business or takeover and integration into a bigger business unit and thus progressive concentration on the supplier level, too. This process of concentration will take place successively on every level of the entire production chain from the top down. Only those with the most financial power will have a chance of remaining independent – neglecting the particular situation of middle-sized companies with no follow-up arrangements in Germany.

5.2.2 Intensified competition due to progressive concentration

For the few remaining manufacturers the concentration at the OEM level leads to a much improved position of power over their suppliers and means that they can directly pass on part of their price pressure. This provokes the creation of countervailing power and thus leads inevitably to a process of alignment and concentration among the 1st tier suppliers and again to an improved position of power here over the next level, which in turn must then adapt its structures to the changed conditions. And so forth!

The result is that because of the worsened sales volume situation in the western industrial countries and the continuing globalization of production locations on the basis of low wages, the entire automotive industry will experience lasting structural change. Declining growth, right down to stagnation, in all the highly developed volume markets will thus lead to a merciless *crowding-out competition* at the manufacturers' level, which is passed on directly to the upstream and downstream producers and service providers.

The progressive decrease in the OEMs' manufacturing penetration and the price and crowding-out competition of the manufacturers will thus have drastic effects on the supply industry, too. The wave of concentration which has been going on at the manufacturers' level for many decades and has now led to a relatively close oligopoly, reached the supply level at the beginning of the 1980's and will cause further companies to disappear from the market in future. With its high density of components suppliers of all kinds, Germany will be particularly hard-hit by this structural change.

The world-wide concentration process in the supply branch is indeed slowing down, but it is still continuing. The current number of 5,500 suppliers will shrink to about 2,800 by 2015 and of the 11 independent automobile companies BMW, DaimlerChrysler, Fiat, Ford, Honda, Hyundai/Kia, PSA Peugeot Citroen, Renault/Nissan, Toyota, and Volkswagen probably only 9 will still be independent.

Fig. 63. Amount of enterprises in automotive industry

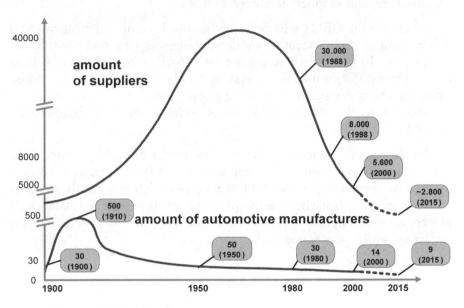

source: Mercer, IWK estimation

This wave of concentration, which starts from the OEM level, has considerable effects on the balance of power between manufacturers and suppliers. The smaller the number of independent manufacturers remaining, and the more these begin to form strategic alliances or intensify their project-related cooperations, the bigger will become the volume of orders placed with the regular suppliers in order to utilize the cost reduction effects of economies of scale. The results are clear: facing the manufacturers on the supply side is a great number of companies which are now in fierce competition with each other for the few but thus larger orders. This imbalance of power also allows the OEMs to increase cost pressure on the suppliers considerably.

Simultaneously the dependence of the regular suppliers on their purchasers increases significantly, since they have to carry a growing utilization risk for their works. In the course of growing order volumes the suppliers have to support high production capacities, in order to be able to meet these orders. If orders are not placed as expected, the supplier can be faced with considerable underutilization which presents an existential threat to earnings and even existence. The supplier can only produce relief by spreading its clientele strategically as wide as possible and trying to become a "regular supplier" for several OEMs.

Conversely the OEMs will be intent on not becoming dependent on a single supplier. The concept of *single sourcing* has very much lost its attractiveness. In their own interest the OEMs will make sure that at least two or three different module or system suppliers are available to choose from. This principle again takes place at every level, so that a stable state of balance is reached when the number of OEMs in the close oligopoly has become consolidated.

This destructive competition for the few orders and the accompanying high utilization risk continues in the course of the concentration process throughout the whole pyramid from the top down. Only when the concentration process has reached the point at all levels, where the balance of power between suppliers and their customers is even, will orders with the necessary profits be possible again.

5.2.3 Excessive demands on middle-class organization structures

The automobile supply segment will grow by 70% by 2015, according to experts' estimations: their added value will rise from currently 417 billion Euros to 700 billion. [59]

In order to cope with this immense growth the suppliers will have to create an additional 3.3 million new jobs by 2015. It will be mainly a matter of qualified jobs, which is shown by the fact that approximately 30 billion euros of additional added value will result from preliminary and serial development.[60]

However, this process will probably not be easy and not all suppliers will cope with it. Pressure of time and costs, an explosive variety of models in all market segments, and an increasing organizational and technical complexity will characterize product development in the whole automobile added value chain. But so also will the race for differentiation by means of innovations, combined with the danger of losing sight of comprehensive technical know-how and the real benefit for the customer. The suppliers

---

[59] Conf. FAST-2015 (2004).
[60] Conf. FAST-2015 (2004).

are the most affected by this. Limited resources of time and capital for each project cause enormous pressure on the suppliers' overall company development processes. Frequently "teething problems", recalls, high guarantee costs, and negative customer surveys are the result of this increasing competitive pressure.

It is just these middle-sized supply companies who are often organizationally overstretched by the structural changes in the OEMs. In the past company organization changed conspicuously at first on the OEM level. Organizational development took its course from the original functional organization through matrix organization to pure project organization for the planning and implementation of development projects.

Then, towards the end of the 90's, a product or series oriented structuring of the whole company established itself in the manufacturing companies as a result of the rising complexity of the individual systems and based on the experience gained from the product oriented structuring of development projects. Meanwhile product oriented company organization has also been taken over by most of the 1st and 2nd tier suppliers. The advantages for the maintenance, application and further development of product-specific know-how are too big to be ignored.

It is foreseeable that because of the lasting trend towards the training of highly specialized problem-solvers product oriented company organization will in future assert itself in smaller supply firms, too – apart from the purely production-technological supply specialists.

The trend in the supply industry towards larger organizational units which are capable of complex organizational solutions is thus indispensable. Those suppliers which cannot adapt their mainly middle-sized company structures to the new challenges of the changed organization conditions in the automotive industry will inevitably be crowded out of the market. This is a big challenge, especially for the management structure of companies with structures typical of the middle-sized supply industry.

5.2.4 Excessive demands on financial power where capitalization is insufficient

The technological and entrepreneurial challenges which face suppliers mean for many of them a considerable *expansion of their financing requirements*. Thus many of the companies are exposed to increasing problems with capital procurement, and above all, loan financing.

This is taking effect just when the entire field of company financing in Germany is undergoing a fundamental transformation. Particularly the very risk oriented lending policy of the banks, formally professed by *Basel II*, is causing altered conditions for the financing of middle-sized companies – which most firms in the automobile supply industry are. A good enterprise rating is therefore the all-important factor for company financing, as it serves not least as a basis for the determination of the interest margin.

The automobile supply companies' *investment structures* have already shifted noticeably in the last few years. While previously the emphasis was on *real investments* in one's own company, in the meantime *direct investments* or investments in research and development have gained much significance. Especially the financing requirements of supply companies for *research and development expenses* have increased greatly. While in the past an advance or at least participatory financing from the purchaser, in most cases an OEM, was often normal, the manufacturers' tense earnings situation has put a widespread end to this practice. For the future hardly any supplier (whatever the turnover size) now reckons with a direct advance reimbursement of R&D expenses from its customer (Fig. 64).

Fig. 64. Decreasing advance financing for R&D

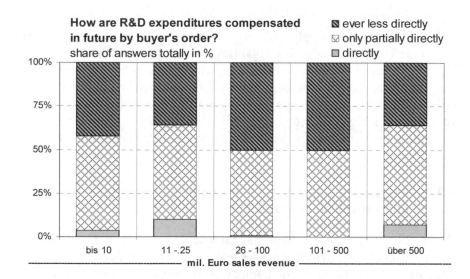

source: IKB

For the supplier, financial hedging, especially of R&D investments for innovations – in view of the high risks of such investments - is subject to particularly high requirements. It is true that the financial institutions have meanwhile developed a series of financing instruments, such as, for example[61]:

- long term bank loan (as a classical financing instrument)
- direct participations (particularly for small and medium-sized enterprises for strengthening their own funds)
- leasing (for fixed assets as an alternative to loan financing)
- cash flow based project financing (for large-scale projects)
- off-balance-financing (e.g. partnership model with the bank)
- asset backed securities (asset securitization)
- borrowers´ note loan (for very creditworthy small and medium-sized enterprises)
- acquisition financing (for the purchase and sale of enterprises / participations).

In the majority of cases however, these methods of financing are not suitable for middle-sized companies which want to maintain their independence from third parties. Nonetheless, the companies must reserve a large part of their disposable own funds and long-term borrowed funds for *project investments*.

5.2.5 Conclusion: decreasing profits, increasing risks, further consolidation

In future the enterprises of the automobile components supply industry will face rising challenges. Due to the outsourcing of production and development activities by the manufacturers a new role allocation will take place throughout the automobile market. The suppliers will be forced to take on new development and production tasks for the OEMs, and to constantly create new innovations, to adapt their world-wide locations increasingly to the changed regional structure of the OEMs, to improve their time-to-market and supplier management, to align their procurement, production and logistics systems securely and stably, etc. Supply firms with mid-

---

[61] For more detailed information on the financing instruments for automobile suppliers see IKB (2003a), p. 37 ff.

dle-sized structure will be increasingly overstretched by all this, at least if they do not hold a monopolistic market niche which allows them certain freedoms.

But not only the usual price pressure from the OEMs will increase. New areas of conflict are arising, which will foreseeably further intensify the margin pressure on the suppliers and their struggle for economic survival. So the OEMs will continue to reduce their (labour-cost-intensive) manufacturing penetration, shift productions and pass on development risks to suppliers. It is a fact that this traditional *"buy or make" decision of the OEMs* has increasingly shifted in recent years to the buy side and will continue to do so in future. Automobile supply firms will have to react to these challenges flexibly, innovatively and above all with cost reductions. For of course every OEM uses outsourcing to pursue the goal of buying products and services externally and thus cheaper than by internal production. Suppliers will have to meet this requirement whatever the case.

With an increasing quantity and complexity of the suppliers' development projects financing requirements increase considerably. This is a dilemma for many suppliers: on the one hand they must raise more and more risk capital for cost and innovative leadership, in order to defend their competitive position with the OEMs, on the other hand the banks are becoming increasingly restrictive when granting loans to supply enterprises especially, or are even withdrawing completely from risk business in this branch. The financial pressure on suppliers is thus rising incessantly. If they then fall back on financial investors, their independence will sooner or later be gone.

While the earnings side is becoming increasingly difficult and complex for the supply firms in the automotive industry the external risks for the companies are also increasing:

- danger of market share losses following the merger of substantial competitors on the global level, with the resulting decline in economies of scale and connected decrease in cost competence and competitiveness,

- intensified procurement competition because of further consolidation on the upstream purchasers' side (concentration wave at OEM level),

- technological innovative jumps (use of electronics, materials), which can lead to an abrupt obsolescence of one's own product programs,

- market entry of new cost-effective competitors from Asia,

- insufficient capital resources for the financing of innovations and growth or for cushioning cyclical sales fluctuations.

But besides all these risks, the global structure change in the automotive industry also offers considerable opportunities for suppliers, which should be made use of with the aid of the right strategic factors.

5.3 Strategic success factors

The task of strategic enterprise management is the identification of competitive advantages and the selection of suitable measures for utilizing and maintaining them. According to Porter's theory of competition a dynamic market is characterized by the interplay of the four determining factors enterprise environment, factor conditions, demand conditions, and related branches, and the two supplementary factors chance and state (compare: appendix 4a, diamond approach according to Porter).

For the competitive efficiency and profitability of a branch, five competitive forces are decisive, which have to be taken into consideration when choosing the right company strategy. These so-called "5 forces" (compare appendix 4b) are

– rivalry of existing competitors in the branch

– threat from new competitors

– threat from substitute products

– negotiating strength of suppliers

– negotiating strength of purchasers

On consideration of the automobile supply industry according to these five criteria, the great extent of the prevailing crowding-out competition can without doubt be anticipated. The negotiating strength of the purchasers (OEMs or module suppliers) is causing increasing competition along the whole added value chain due to the concentration process described above. The selection of the correct success strategy is therefore decisive for survival in this merciless crowding-out competition in the automotive industry.

5.3.1 Innovation strategy

Innovations are a decisive factor for gaining competitive advantages. It is not a new discovery, that innovative enterprises as a rule have a strategic advantage over their competitors. It is only by constant innovation that a company can secure a technological lead in highly competitive markets, and thus has the possibility of at least temporarily and within limits, withdrawing from the price competition and making *pioneer profits.*

Especially in the automotive industry with its current crowding-out and concentration process innovative skill, or still better, innovative leadership is of existential significance:

• On the one hand for the manufacturers, which have to differentiate their model range from the competition and spur on the buyer readiness of a saturated clientele again and again.

• On the other hand for the suppliers, which have increasingly replaced the manufacturers as innovators in the past 20 years, because the focus of innovation has shifted more and more away from the basic innovations typical of the manufacturers (brakes, gears, chassis, engine management systems, propulsion technology) to the vehicle periphery (comfort, air conditioning, communications, entertainment, etc.); that is, fields which have nothing to do "with oil and gasoline".

Fig. 65. Innovation strategy

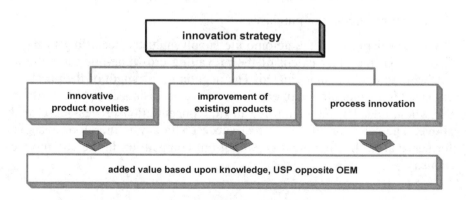

source: IWK presentation

Since the invention of the automobile more than 100 years ago, the series of the most varied innovations around product, process and branch has found no end. While in the early days of the car, basic innovations were often determined by chance and linked to the persons of ingenious developers and amateur constructors, innovation research is today done systematically at progressively increasing expense. Suppliers can choose between various strategic options here.

The aspiration to innovation leadership is decisive for competition for all suppliers alike, whatever their specific product program and their closeness to the OEM within the value added chain. For innovations are complex and relate, contrary to popular opinion, not only to the product but just as much to the production process, materials, company organization and production site.

The major part of automobile components supply enterprises in Germany is thoroughly aware of the significance of innovations for their own competitiveness and attach great significance to the topic complex "innovations". In a survey by ZEW it was shown that innovations in vehicle construction are generally considered more important than in other manufacturing branches. In vehicle construction it is above all innovation activity in new technologies related to the product automobile which are given great importance, while process innovations and individual customer solutions come off well under the valuation of the other manufacturing branches. This would seem to show that in these areas the innovation willingness of the automobile suppliers is merely limping along. A successful company will have to expand its innovative activities above all in this area, if it wants to gain an edge on its competitors.

Concerns that the potential for new innovative ideas and technological solutions in the automotive industry might one day be exhausted are superfluous. Just because the automotive industry is highly engineered, complex and extremely globalized, still more: because it is in the broadest sense a mirror image of the constantly changing economic, technical and social environment, it is a constant source of innovations with completely different characteristics. all suppliers – whatever their shade – therefore have to face the necessity of innovations.

This also applies to producers of standard components and supply parts. Even if they have comparatively little possibility to cushion price pressure by means of technical innovations in the function of the products itself, and thus escape the compulsion to high volume, they do still have the possibility to innovate process, materials and site. Especially in times of heightened efforts on the part of all OEMs to reduce the weight of vehicles material innovation is an important competitive factor.

Fig. 66. High significance of innovations in vehicle construction

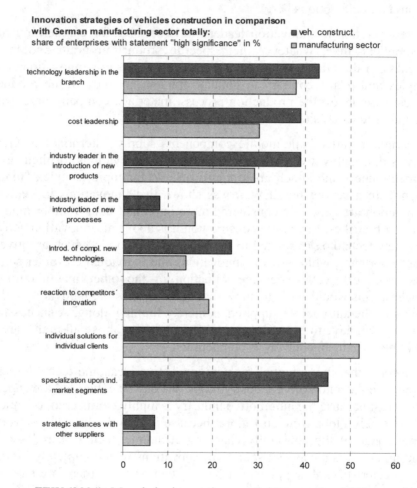

source: ZEW (2004): Mannheim innovation panel, interrogation 2003.

Specialists who develop and construct the complex and high-grade components necessarily must aspire to innovation leadership, because their customers generally expect/demand it of them. This gives the specialist the chance, however, of persuading his purchaser to pay a top price with which he can work profitably.

If the *product innovation strategy* is followed, two dangers must be avoided.

- However significant innovative products and solutions are, quality and reliability of the components must still not suffer. Poorly conceived products do not only annoy the customer, they are also a cause of high costs. And that not only because of the necessary rework, but above all because of the loss of image and customer/purchaser confidence. Especially the increasing use of electronic systems in cars, with unsolved interface problems, means that uncoordinated parts or systems go into large-scale production too early. This is mainly because of the increased price and deadline pressure, which leaves the developers too little time for testing. Particularly when parts are increasingly more complex and temperamental, as is the case in the field of electronics and software, quality assurance should still have precedence over innovation leadership.

- "Fictitious innovations" should be avoided. Product innovations must be directed to the needs of the customer and contribute to the strategic brand value of the company. Innovative products in the components supply industry often lack clear market and customer orientation. In particular the European companies have a strong emphasis on engineering, are driven by technology and frequently have too little market and customer feeling. Although technically brilliant, innovations have often no relation to customer needs. Starting from company targets and product image the automotive supplier should define clear, realistic goals and from there include its entire development process in its innovation strategy.

Enterprises which can cope with the balancing act between customer or market oriented innovations and economically interesting solutions are the ones which are successful. The key to success is the strict alignment to the client's needs. Whereby for a company in the supply branch it is not only the end customer who counts as a client, but primarily its immediate purchaser in the added value chain. This means that innovations must be aligned to all the players involved along the added value chain, to ultimately be successful in the market. To this purpose a supplier needs exact information from all the upstream direct and indirect target groups, from the actual purchaser through to the car driver as final customer. Supply market research and supply marketing will therefore become more significant in future.

5.3.2 Cost reduction strategy: increase in productivity or transfer to low-cost countries

An essential task for every supplier is the consistent implementation of its cost optimization strategy. In intense competition, as prevails in the supply market, no enterprise can allow itself to leave available cost-saving potentials unused. While for a whole decade it was almost exclusively the suppliers which were at the focus of the OEMs' cost reduction strategy (Ignatio Lopez is a synonym for this policy), the necessity to optimize has now reached the manufacturers themselves: reorganization and massive cuts in labour costs by job reduction at the OEMs govern the picture at present.

However, this has not resulted in any decrease in the price pressure on the suppliers. They have several cost-optimizing options to choose from (Fig. 67).

Fig. 67. Cost optimization strategy

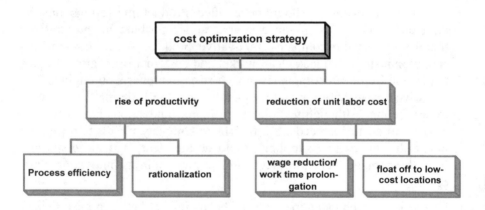

source: IWK presentation

In principle, there are two levers for optimizing costs. One is the improvement of process productivity. This can take place either by means of rationalization or more efficient process structuring. This includes for example a more efficient use of materials by reducing the wastage or the use of machines with a higher rate of production.

The German automotive industry has modernized its product range considerably since the end of the 90's. The proportion of sales volume with innovations dropped only slightly in vehicle construction to 11.3%. In a branch comparison it is still above average (industry total: 7.6%). But the proportion of cost-reduction achieved by innovative processes has further decreased. The companies involved in vehicle construction were able to reduce their average unit costs by only 5% from process innovations in 2002. This was the lowest value since 1994 (Fig. 68).

Fig. 68. Sales shares with innovations and cost reduction in vehicle construction

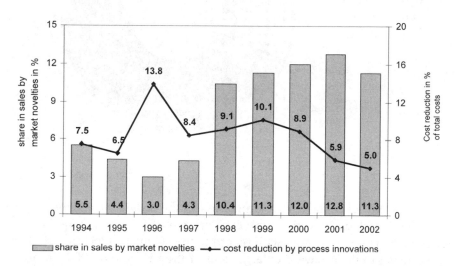

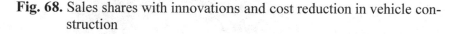

share in sales by market novelties ◆ cost reduction by process innovations

Reading help: in the year 2002 the sales volume share, produced by the enterprises of the branch with market novelties, amounted to 11.3%. By process innovations the enterprises reduced their unit cost for 5%.

source: ZEW (2004): Mannheim innovation panel, interrogation 2003.

But appearances are deceptive: process innovations in German vehicle construction are aimed less at the pure reduction of costs (11%) than at a combination, with quality assurance playing a bigger role. In 77% of all process innovations the improvement of quality played a role and in more than half (52%) they were accompanied by cost reductions, which reflects the quality orientation of the German automotive industry.

However, 12% of the process innovations were neither cost nor quality oriented, but served other goals, for example the improvement of production flexibility (Fig. 69).

Fig. 69. Effects of process innovations in vehicle construction (2002)

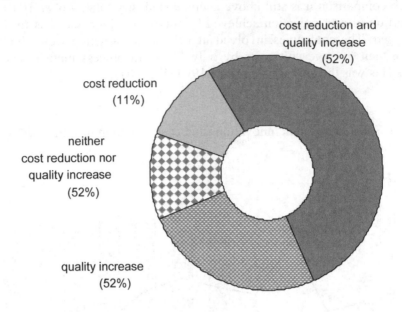

cost reduction and
quality increase
(52%)

cost reduction
(11%)

neither
cost reduction nor
quality increase
(52%)

quality increase
(52%)

source: ZEW 2004

The second lever for cost optimization is the *reduction of unit labour costs*. This problem is very acute, especially in Germany as a high-wage location, and especially, too, with regard to the heightened location competition owing to the eastward enlargement of the EU in mid-2004. In order to be able to keep up with the low-wage countries in the prevailing cost competition, supply companies at existing production locations theoretically have the possibility of bringing costs down to a competitive level by wage cuts for employees. In practice however, the reduction of direct income components is only very rarely enforceable and limited at best to voluntary company payments (holiday pay, Christmas bonus, etc.). If one wants to protect macroeconomic purchasing power as much as possible, the lengthening of working hours without compensatory wage increases is a better alternative, which in the end produces a reduction of the effective hourly wage, and thus of unit labour costs.

This instrument has been employed increasingly since 2004 in the German automotive industry, to avoid job displacements which as such are necessary, and in agreement with the workforces concerned.

This shows the insight on the part of all those involved in the automotive industry that high wages are one of Germany's biggest problems as a location in international competition, and will inevitably lead to a massive displacement of jobs abroad, if nothing is done to counteract it.

But because working hours cannot be lengthened endlessly, established industrial countries, and above all Germany with its high level of industry, will have to come to terms with the fact that companies will increasingly reduce their costs by shifting their wage-intensive production to low-cost countries. Especially where less complex standard components are concerned the volume suppliers will be able to assert themselves, on the one hand fully exploiting the effects of scale and on the other hand taking all sorts of measures to minimize costs. Besides rationalization measures and increases in efficiency at existing production sites, this includes above all the exploitation of location arbitrage by shifting production to low-cost countries, chiefly to Eastern Europe.

The labour costs tend to decrease, the further east one goes. On the other hand, logistics and transport costs increase with distance. Thus it is important, especially for the supply firms, which make most of their products for purchasers in Germany or Western Europe, to evaluate the trade-off between low labour costs and high logistics costs correctly. Depending on the typology of a supplier and the special characteristics of the components produced, each supplier will have a different geographical profitability limit, up to which a shift would still make sense. For the world's second-biggest supply enterprise, Delphi, this profitability limit is almost 3,000 km from Germany, according to the company itself, in the Russian town of Samara, where the company produces cable sets for purchasers throughout Europe (see Fig. 70).

In addition the Asian low-wage countries such as China, which are considered to be the growth markets of the future for automobile sales, are already being focussed upon as production locations for western automotive producers and suppliers. A shift of production is worthwhile here, not only because of the low production costs, but also because of increasing sales and demand locally.

The results are not difficult to estimate: the more production sites the OEMs build in low-cost countries, the stronger becomes the pressure on the regular suppliers to also become active for their local purchasers in these regions. Having once gained – positive – experience locally, the floodgates are opened for subsequent production shifts, to convert the cost advantage into a competitional advantage in the western sales markets.

Fig. 70. Profitability limit for production shifts

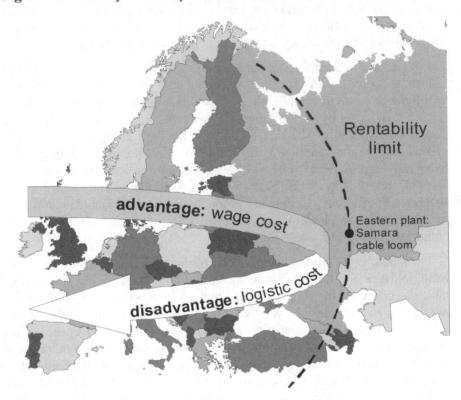

source: Delphi (2005)

Even if some politicians in Germany call these measures unpatriotic and "betrayal of the homeland", it is nothing other than the necessary exploitation of location advantages in order to be able to survive in tough global competition. Or, as Nobel Peace Prize winner, Milton Friedman, already expressed it in 1970, "the social responsibility of business is to increase its profits"[62].

It is true that the VDA states that the previous rule of thumb, according to which every third job created abroad creates another in Germany, is now no longer valid and instead only means that a job in Germany will be kept. "Presence and growth in low-wage countries safeguard jobs in the domes-

---

[62] New York Times Magazine, September 13, 1970

tic economy"[63]. However, the VDA also states it should not be overlooked that over 40% of the export value of the German automotive industry already comes from upstream supplies from low-wage countries, and the tendency is increasing. These reorganization and cost-reduction activities on the part of the manufacturers and suppliers are offensive steps to increase competitiveness and keep more value creation at the location. VDA president Gottschalk: "The walk is not yet finished".[64]

As already shown, it is vital for a company in the fierce market of the automotive supply industry to gain competitive advantages over the other competitors. Where true innovations are concerned, costs and price do not play a big role for the purchaser. That is different where simple technical components are concerned. Low prices are the main competitive factor for the supplier here, considering the high price pressure on manufacturers. And with labour-intensive components these low costs are easiest to reach in low-wage countries.

Fig. 71. Differing shifting pressure in the supply branch

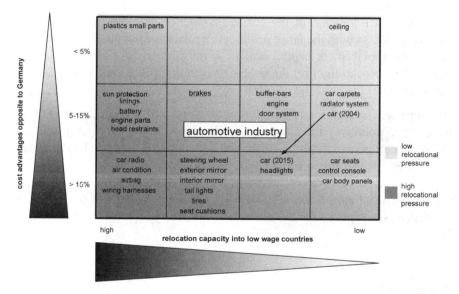

source: BCG

---

[63] VDA-President Gottschalk on the VDA-annual press conference (2004-01-29).
[64] In the same place.

According to a study by the Boston Consulting Group[65], by 2015 up to 50% of the parts for a car made in Germany will come from abroad. The pressure to move is particularly high for suppliers of standardized systems and modules such as steering wheels, air conditioning and wiring harnesses (see Fig. 71). These can be made in mass production, but still include a high proportion of labour costs. Thus a shift of production to low-wage countries is particularly suitable for reducing costs without endangering the quality of the product. As for example in the case of the Leoni Plc.[66]

Primary parts, such as for example injection-molded small parts, can however usually be produced fully automatically, so that a relocation of production (even if it was relatively easy to do in principle) would in this field only make very little difference to costs. On the other hand, close proximity to the OEM plant is of primary importance for suppliers which produce complex and individualized systems such as dashboards and seats. By means of better infrastructure and optimized logistics management the just-in-time and just-in-sequence requirements of the OEMs will be easy to fulfil from nearby foreign countries (e.g. Eastern Europe). The pressure to relocate will thus increase in the fields which until now have been spared to a great extent.

However a relocation of the entire production is often not worthwhile for a supplier, as with the large producer, because of the costs connected to it. But the replacement capital and new investments are already flowing abroad; the relocation of production is thus taking place creepingly, but the long-term effect is none the smaller.

The suppliers' research and development work will stay in the traditional automobile countries for now, as will the production of high-grade and complex components, where high quality and knowledge standards are guaranteed. But in this area, too, the established western industrial countries are finding themselves increasingly in a rising competition with the newly industrializing countries. A comprehensive analysis of the future of Germany as a location for the automotive industry follows in chapter 6 of this study.

---

[65] BCG (2004).

[66] The Leoni Plc., a supplier of vehicle electrical systems and cables, for example still occupy 10 % of their 30,000 employees in Germany and 90 % in (mainly Eastern European) low-wage countries. The staff costs for these 27,000 employees abroad is in total as high as the costs for the 3,000 employees in Germany. Currently Leoni is planning to relocate a plant from Hungary into the Ukraine, for in Hungary also labor costs have in the meantime become too high in international competition.

5.3.3 Expansion strategy: more activities within the added value chain

In order to safeguard profitability suppliers will have to produce more cheaply or be more innovative in future. The guiding principle here is the perusal of a *content strategy*. At the heart of this is the safeguarding of profitable enterprise growth by increasing one's own added value share in the end product, the car. Thus it is about taking over more added value from the OEM.

As described above, manufacturers will in future hand over more and more production and development activities with above average marginal costs to the supply companies. This shifting process demands on the one hand maximum enterprise-related skills on the part of the supplier, such as complexity management of the overall process and the coordination of the various interfaces. On the other hand its market position in relation to the OEM is of course improved.

The system and module specialists will also profit from this withdrawal of the OEMs and will be able to improve their competitive position, as will the development service providers.

The structural change in the automotive industry thus offers an excellent opportunity for growth oriented components supply companies to expand their activities strategically within the added value chain, on the basis of the increasing reallocation of roles between manufacturers and suppliers.

Fig. 72. Content strategy

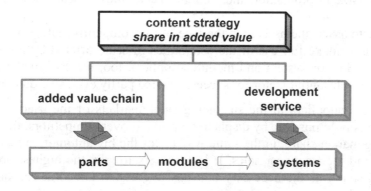

source: IWK presentation

This presupposes a change in the suppliers' thinking. In the past they were predominantly concentrated on the production of parts by order and to the exact specifications of the client. They are now challenged in the course of inter-company cooperation in the sense of strategic partnerships with pronounced development, production and logistics competence, and no longer as an extended workbench. The automotive suppliers are taking over more and more of the OEMs' tasks. This applies to development, and the production right up to taking over integral and complete delivery chain management.

The German automotive manufacturers' added value penetration will continue to decrease in future for the reasons already described, and the added value penetration – in scope and responsibility - of the suppliers will rise accordingly (see Fig. 56, p. 123). A decisive competitive factor for the manufacturers will become to concentrate increasingly on core competences and strengths and not to "lose themselves" in activities in unrelated industries. Lateral diversification, which was preached as the guarantee for success in the 80's by business consultancies of high repute, and which with one accord all the automotive companies were taken in by, has long proved to be the wrong path.

5.3.4 Growth strategy: economies of scale and cost synergies

Closely related to the strategic extension of activities within the added value chain in the sense of "more of the new things" is the volume strategy in the sense of "more of the old things". This is about exploiting advantages of size in production and attaining economies of scale. Above all in the manufacture of complex standard components only those suppliers will be able to assert themselves in the face of the competitive situation, which are able to make full use of the economies of scale offered by mass production. But for system and module suppliers, too, the size of the enterprise is a substantial factor for success, indeed partly even sine qua non.

For the suppliers in the stagnating sales markets of the triad, volume growth is only possible by displacing or taking over competitors, that is by gaining market share, in the same way as for the big automotive manufacturers and their new car sales. It is necessary to gain the highest possible market share, as the necessary unit cost digression aimed at can only be achieved by mass production. Concentration of the market is the inevitable result.

Fig. 73. Volume strategy

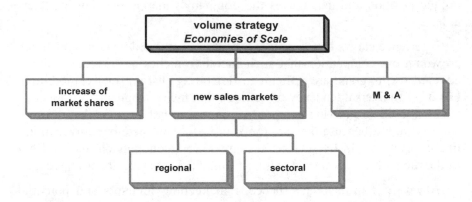

source: IWK presentation

Apart from in the existing markets by market share gains an extension of production volume is also possible by gaining new customers in new sales markets. This can take place on the one hand *regionally*, whereby it is becoming apparent that in this field the markets "in the old world" are already allocated and the competition here is characterized by concentration rather than by expansion of the market. Any regional expansion will thus take place mainly in the "new markets", for example the BRIC states, with the opportunities and risks described above, which will have to be taken into consideration.

On the other hand, there is also the possibility of increasing volume by a sectoral expansion into new markets. Thus successful suppliers manage to gain structurally new customers for their products again and again from unrelated areas of industry. Examples of this are the entry of Bosch into the heat technology market or WOCO into the cable-guided transport sector.

5.3.5 Niche strategy: specialization

A fundamentally different strategy for a supply company is the specialization on a particular niche, to be able to meet particularly sophisticated customer requirements there and operate highly profitably. The prerequisite for this is however always a virtually monopolistic position (for example on the basis of patents or special production know-how), which leads to a unique position and thus makes the company's market position invulnerable.

Experience shows that it is just in times of saturated markets and fierce crowding-out competition that such lucrative niches never remain undiscovered for long and are difficult to defend against outsider competition. For it is a universal tendency of markets that niche suppliers are sooner or later exposed to growing competitive pressure, not because they are not successful, but because they are too successful and therefore attract imitative competition. In free-market systems every monopoly cheats itself out of the roots of its success in the long-term, unless it is legally arranged.

However, if the supplier succeeds in keeping imitators and potential market entry candidates at a distance by strategic means such as exclusive customer relations, high quality and special know-how, the future offers lucrative niches in the automotive industry for specialized supply companies. The list of examples is long.

The automotive manufacturers try ever harder to cater to the individual wishes of car buyers, in order to generate further growth. As a result of the growing variety of models and market segmentation, not all components can be made of identical parts and platforms; there will always have to be a certain proportion of special solutions, too. Suppliers can use these niches, where as a rule not large volume, but small production series are required, for specialization and there achieve high profitability. Empirical surveys by the German Industry Bank [67] (IKB) show clearly that as far as profitability is concerned, "small is beautiful".

However, it must be conceded that a niche is not necessarily connected with small market volume. The example of Microsoft shows that market niches can very well become large, one succeeds in screening them against outsider competition or if they are not noticed by potential competitors for a long time. But it is probably mostly the case that successful niche suppliers are followed very quickly by imitators, which break up the niche as a market by their entry.

---

[67] See IKB (2004).

Fig. 74. Suppliers´ yields by size

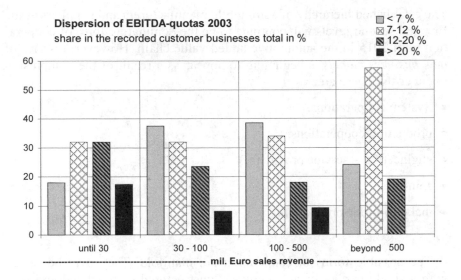

source: IKB

For a niche supplier the direct customer relationship with the OEM is above all decisive. The enterprise must specialize in special customer solutions and uphold a constant exchange of information with its purchaser. The demands on the niche supplier at this point are high, it must possess absolute innovation leadership in its field and be in a position to react to special customer wishes. However, the niche business can be profitable with small unit figures despite high innovation expenses, if the product is brand-forming for the manufacturer and if it can be marketed with temporary exclusivity. Then the OEM will be prepared to pay a top price for it. Usually it is just these special solutions from the niche supplier which enable the manufacturer or system integrator to create a contrast between its product and the competition.

The specialized supplier if innovation thus makes a decisive contribution to its purchaser's brand attractiveness. This results not only in a close customer bond between the final customer and the automobile manufacturer, but also of course between the supplier and its customer, the manufacturer. The niche supplier is thus not as easily exchangeable as other suppliers, and has a better negotiating position with the manufacturer. As long as the manufacturer is dependent on the creativity of the niche supplier it can operate at sufficient profit. This is shown by numerous examples.

5.3.6 Cooperation strategy: creation of clusters

The established hierarchy of automobile manufacturers and suppliers at the first and second levels will continue to be the dominating form of cooperation until 2015 in the automotive added value chain. However, a series of *new business models* is beginning to appear as a result of the compulsion to cut costs. These are:

- System cooperations,

- production cooperations,

- engineering – service providers,

- spin-offs,

- make-to-order-production.

The possible forms of cooperation are manifold. They may be set up with a single partner or as a network, unstructured or regulated and may operate by means of supply and production contracts, licenses or joint ventures. The formation of networks related to only one project is also possible. Cooperations therefore save the resources of the individual companies, are low risk and can be adapted quickly to changed conditions. According to experts this will change cooperation considerably in the whole branch. A study by Mercer[68] identifies more than 20 new forms of cooperation, by which a new quality in cooperation between automotive manufacturers, suppliers and service providers can be attained.

If these new forms of cooperation are used consistently and mutually, they promise lower costs and considerably improved profits compared with the traditional forms of cooperation: on average over the last 10 years the EBIT margin of the automotive manufacturers was 4.8% and of the top suppliers 6.5%. The mega trends described in chapter 3 suggest that if these traditional forms of cooperation are continued, the margins of all those involved will come seriously under pressure. Because of the growing crowding-out competition among the automotive suppliers, above all pressure to come up with innovative products and the connected investment risks will rise considerably for the suppliers. The individual supply firm will hardly be in a position to produce the development performance necessary for complex application solutions alone in future. Sufficient syner-

---

[68] Conf. FAST-2015 (2004).

gies to meet the increased demands in global competition, and with sufficient profit margins, can only be generated by cooperation.

Innovative business models combined with a new quality of cooperation will thus counteract the general erosion of margins. Taking the results of the Mercer study as a basis, savings in costs of between 600 and 1,000 euros per vehicle can be made in the cooperative business model for the branch. This means car makers and suppliers can achieve a 3 per cent better EBIT margin and capital returns (ROCE) can rise by between 4 and 10 per cent.[69]

The shift of development activities and responsibility to the suppliers has been leading to a re-positioning of all participators in the automotive product generating process for some time now. In particular cooperations between manufacturers and module or system suppliers have become standard in the last few years. In this way the manufacturers pursue several targets at once, among others, the reduction of manufacturing and development penetration, standardization, reduction of work in progress and an increase in development efficiency.

The development and production of a complete module or system cannot normally be carried out by a large supply company alone. For one thing, this is because of the economic risks of such projects. The other thing is that the technical content of the various modules or systems with their variety of functions and materials usually exceeds the competence spectrum of a single supplier. Add to this the increasing networking of individual systems with one another (particularly in electronic systems). The result is the formation of development cooperations, with the inclusion of specialists, whereby the classic hierarchical cooperation forms are being increasingly broken down. Instead *development networks* are increasingly asserting themselves on the supplier side, in which all those involved are equal, independent of their size. There are still market potentials in the fields of calculation and simulation, in particular for development service providers.

No enterprise has to invent the wheel a second time. Ideas from different companies can be combined so that from two small ideas one big innovation can develop. Ideas from one company can inspire further ideas in another company; the view for new aspects can be opened up. Cooperation and the forming of teams beyond the limits of the company can open up new potentials. In short: the motto "together we are strong" gains a com-

---

[69] Mercer (2003).

pletely new meaning for small and middle-sized supply companies with regard to their strategic positioning and sustainability. The chance of being able to hold one's ground in the market in the long-term increases at the same rate with which they dispose of a higher mutual professional know-how, higher innovative power and better capital funding.

In recent years development cooperations are also being pushed at the *manufacturers' level*. Thus the capital-intensive development of engines is increasingly taking place as a manufacturer spanning activity. But whole platforms, too, are increasingly being developed in cooperation between manufacturers to reduce costs, as shown for example by the joint development of the platform for the VW Touareg / Porsche Cayenne (Fig. 75). In future such manufacturer spanning development cooperations will increase, which will lead to higher demands on coordination.

Fig. 75. Increase of development cooperation

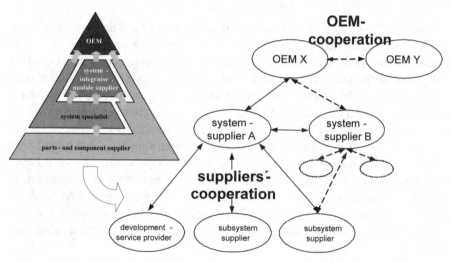

source: Baika, IWK presentation

Due to specialization, cooperations will present a heterogeneous picture with regard to the size of the cooperation partners and their structures. Besides the hard cooperative criteria such as cost reduction, shortening of time to market, improvement of results, etc, a substantial success factor for development cooperations today is to be found in the soft criteria.

This means above all personnel and cultural differences or views between the cooperation partners. Interpersonal incompatibilities in these soft factors are one of the main reasons for the failure of many cooperations. Above all the top management level must demonstrate the will to open partnership and learn the lessons from the many failures, in order to avoid the discrepancies, which are mostly very personal. Particularly at the level of the middle-sized suppliers the "clash of cultures" between the acting personalities is the most frequent cause of disappointments and the failure of cooperations.

The formation of cooperations, especially development and system clusters, is gaining importance with regard to the balance of power in the automotive industry, as they contribute substantially to achieving an improved position of power for the suppliers involved. By having the technical lead and supplying complete systems several small companies can also gain an improved negotiating position in relation to their purchaser. While teamwork and cooperation in the field of development, both between OEMs and suppliers and between the suppliers themselves, are meanwhile well advanced and functioning well, cooperations in the fields of purchasing and production, which today are not very well developed, will in future be noticeably extended. Improved negotiating positions and thus cost advantages can be achieved in purchasing particularly, by the formation of partnerships and cooperations. These advantages must be exploited, to be able to survive in the market.

In manufacturing, suppliers will expand assembly penetration in future, to supply the client with entire modules instead of individual components. This takes place either by reorganization, integration of company-produced components or by strategic purchasing, partnerships, and / or joint ventures.

For reaching strategic goals, cooperations among equals, in comparison to mergers or takeovers, represent the most flexible and least risky solution. It is decisive for success at this point that each of the partners knows ahead of negotiations what resources and skills he has at his disposal, whether the project corresponds to the company's own long-term goals and whether it is realistically practicable with the chosen partner.

5.3.7 Location strategy: go to where the OEMs are

The global extension of suppliers' production activities is nothing new and does not always take place purely for cost reasons, Thus successful German supply companies, for example Bosch, Brose, EDAG, WOCO, etc, have already been actively present abroad for more than 20 years with their own development and production facilities, and have opened up new markets very successfully by supplying the automotive manufacturers there. In doing this they followed an independent globalization strategy long before certain OEMs, and before the term was even coined.

In the meantime the German automotive groups have increasingly discovered important sales markets as production locations, as Volkswagen in Brazil and Mexico, BMW and DaimlerChrysler in the USA and South Africa; and all of them together in China. Ford and GM, for their part, are part of an internationalization strategy as already practised by their parent companies before World War II. In many countries in which complete production lines are not worthwhile, CKD plants[70] have been erected.

This global orientation of the manufacturers has resulted in the passive globalization of many middle-sized suppliers. In the course of their own globalization the manufacturers are increasingly demanding a worldwide presence of their direct business partners and the building of their suppliers' production capacities within reach of their own plants. But passive globalization has an advantage over *active* globalization in that there is already a fixed purchaser at site. The process of globalization and concentration at manufacturer level thus directly affects the location policy of many independent supply companies. The necessity of supplying a manufacturer at all production locations worldwide with parts and components demands knowledge and international presence of the supplier.

It is obvious that ever more suppliers are following their clients and building additional production capacities abroad. The procedure differs depending on the starting position. Production begin may take place either by takeovers or joint ventures with a local company, or by the company's own investment in plant and equipment. The first possibility offers the advantage of taking over an existing network at site and not having to

---

[70] CKD = Completely Knocked Down, i. e. in the plants the vehicles are manufactured from parts sets, which are gained by the dissembly of complete vehicles. This technique is suitable if at the plant location there is no manufacturing possibility for the necessary parts. Often this technique is also employed for customs reasons, to avoid high import duties on finished vehicles.

build everything up themselves, which is usually a protracted and risky process. Cooperation projects, joint ventures or mutual enterprise holdings may help to improve the strategic position of a company here, but they also hold risks, for example from know-how outflow, which must be assessed in each case. China is a brilliant example of this. As a rule one of the joint venture partners comes off worse.

The high and risky capital expenses are a stumbling block for the necessary globalization of a company, as it is always connected to investments for setting up the company's own production facilities abroad. This presents big problems above all for the smaller of the middle-sized firms. Without sufficient capital resources they cannot expand abroad. But just this will be increasingly necessary for survival in the market in future. Further concentration is the inevitable consequence.

A strategically important aspect of this development is that the number of plants is to date growing faster than sales or production volume, that is, overcapacities are being created. It is true that this can never be completely avoided, but as far as possible factories should be planned such that they can operate profitably even at partial utilization, amortize quickly, but are definitely flexible and adaptable enough to be able to adapt quickly to new production requirements.

A further important aspect for suppliers when making decisions about OEM-driven investments abroad is not only whether, but also for which manufacturer or which strategic alliance they should decide. They need purchasers which can guarantee a certain utilization of the new production facilities. If one applies the rating results from chapter 4 the suppliers will have difficulty with investments in favour of some of the manufacturers.

5.3.8 Financing strategy: securing increasing capital demand

Rising innovation and cost pressure on the supplier side mean that increasing, if not the most, significance is given to financial power and capital resources of the companies. If a company can no longer raise the heightened capital requirements for location optimization, volume increase and innovative research, withdrawal from the market is the only solution, either by discontinuance or acquisition by a competitor.

Innovations in particular usually demand long drawn-out development times with correspondingly high R&D expenses and thus financial "staying power". Advance financing from the automotive manufacturers is taking place to a decreasing extent, as these are increasingly having to struggle with sales problems of their own. Added to that is the considerable rise in financial requirements due to the expensive and risky expansion of the suppliers into the growth regions of the automotive industry. In the course of globalization middle-sized suppliers have to be present in Asia, Eastern Europe and Latin America, since the manufacturers demand it urgently from their suppliers and they can often only remain competitive because of the cheaper production in these countries.

Especially the smaller supply companies have to struggle with high investment costs. According to an analysis by IKB[71] the investment quota, related to sales revenue, for suppliers with less than 30 million euros in sales revenue in Germany in the last few years was well above that of the companies with stronger sales figures, and above all, well above the cash flow quota, in contrast to the bigger firms. In 2003 the investment quota of the smaller suppliers was at over 14% more than twice as high as that of the companies in all other size classifications.

Fig. 76. Propensity to invest in the automobile components supply industry 2001-2003 in relation to sales revenue

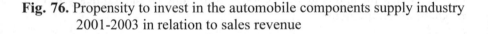
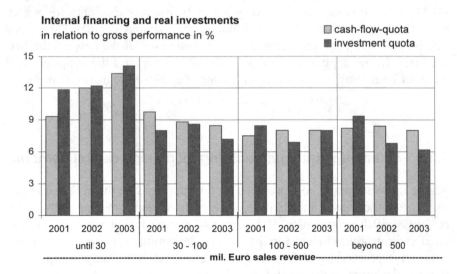

source: IKB

---

[71] IKB (2004).

In addition the automotive manufacturers have been increasing the price pressure on their suppliers, which is then passed on along the added value chain so that is often the middle-sized companies of the branch which are in part under significant threat due to their insufficient capital resources and high level of debt. The new regulations for equity capitalization from credit institutions *(Basel II)* will also have considerable influence on the creditworthiness of just these middle-sized enterprises in the supply industry, via higher requirements for credit rating and the openness of borrowers. [72]

Basel II urgently demands a *rating procedure* for the future of banks, with which borrowers would have to be individually rated. Both enterprise code figures, such as equity value or solvency ratios, and the so-called soft facts, such as the personality of the entrepreneur or risk management, play a role here. The evaluation of these figures and information will not only be considered in relation to the past, but also in relation to the future. As a result the previously applied standardized creditworthiness weights will be disregarded. The credit institutes will underlay the credits approved with varying equity ratios, the minimum of which will depend on the creditworthiness of the borrower and the quality of the lending collateral, in short, on the company's rating. Companies with bad rating results will - if they can get credit at all – have to bear higher credit costs in future, those with good results will get more reasonable terms.

There are some arguments that Basel II could have a negative effect on the credit supply for middle-sized enterprises, on the basis of the suggestions currently being discussed, whereby the increasing capital requirements gain explosiveness as a bottleneck criterion. It is especially significant that the equity ratio in particular will play an even more important role in the assessment of creditworthiness in future. Large parts of the German small and middle-sized businesses, as a rule family owned firms, unincorporated businesses and public companies, with their traditionally low capitalization, will find it increasingly difficult to obtain loans at reasonable terms. This will, incidentally, also apply to output loans. Many middle-sized firms will thus be forced to develop a stronger orientation to

---

[72] The consultation paper aims at a revision of equity capital regulation for banks. It is based on a "three-columns-model". The first column covers the minimum requirements for equity capitalization, the second column covers the bank regulatory supervising process and the third column market discipline and the enlarged duties of disclosure. See on this the IWK study (2001): "Rating as a challenge for small and medium-sized businesses and banks –Basel II and its effects".

the capital market. For further and more detailed information please refer to the relevant literature[73].

It must be noted that investment costs, above all for smaller firms, will rise steeply and the increasing capital requirements can become a bottleneck criterion for many middle-sized supply companies. Also because Basel II will tighten lending to middle-sized firms if they do not try hard to gain a higher equity position and a better rating. Unless the enterprises take measures, Basel II will lead to expensive loans, the increasing capital requirements for safeguarding an independent market presence will not be able to be covered, the number of insolvencies will increase and / or the concentration process in the supply industry will continue.

5.3.9 Conclusion: profitable growth is possible!

Just as the OEMs themselves, the enterprises of the automotive supply industry also face further increasing challenges and burdens. Because of the outsourcing of production and development activities by the manufacturers a new allocation of roles is taking place in the whole automotive industry. The suppliers are compelled to settle their locations close to the OEMs worldwide, to improve their time-to-market and supply management, to adjust their procurement, production and logistics systems securely and stably, and above all to tap new cost cutting potentials again and again. The alternative to all that is to withdraw from the market.

Falling earnings and rising risks for the marginal suppliers involve the further advance of the concentration process among the automotive suppliers. For the increasing cost pressure will in most areas of the branch only be bearable by attainment of influential market size – and thus market power. As a result, mainly, of the further tightening of the manufacturers' oligopoly, also in the form of conglomerates (development and production cooperations), the space for suppliers will shrink progressively, because the number of customers / purchasers will decrease further. Therefore every supplier must set itself the strategic goal of being present with its products as far as possible in each of these manufacturer conglomerates. If that succeeds it can operate profitably independent of the OEMs' crowding-out competition. Because the decline in orders from one manufacturer will be balanced by the increase from another, at least as long as the mar-

---

[73] IWK (2001), IKB (2003).

ket as a whole does not shrink. And this is not really an imminent danger in the automotive market, despite individual sales problems.

The suppliers thus find themselves in a strategically much better starting position than the OEMs. They "only" need to succeed in not becoming dependent on one single manufacturer or manufacturer conglomerate, and in being able to supply various purchasers. The motto for the supply branch is therefore: "Being there is everything".

The relative size of an enterprise in its market or the attainment of an internationally relevant market share in its field of activities is therefore the decisive survival criterion. As a rule of thumb, belonging to the top 3 on the home continent and the top 5 in the world can be applied.

Brake systems are a very good example of this relevance of market share. Worldwide Bosch and Continental Teves are the two enterprises which dominate almost two thirds of the market. While these two suppliers have divided the European market between themselves, two other enterprises are active in North America, which play a relevant role there. In Asia the market is still somewhat more heterogeneous, however, only one Asian supplier achieves a worldwide market share of over 3 per cent (Fig. 77, Fig. 78).

Fig. 77. Market shares of brake systems, world-wide

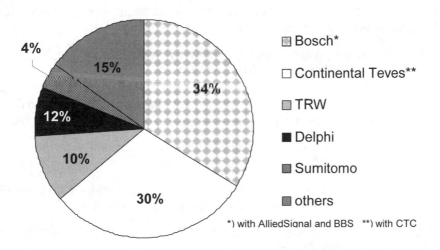

source: Automobil Produktion

Fig. 78. Regional distribution of market shares for brake systems

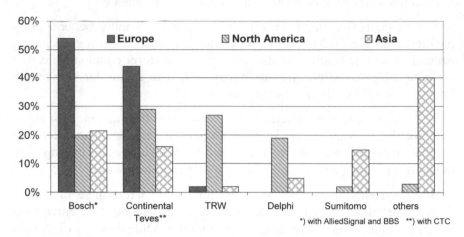

source: Automobil Produktion

In the course of globalization the enterprises will increasingly build up their activities worldwide and crowd out or take over the smaller, regionally based suppliers even more.

This example of the market share of the 1st tier suppliers in the field of brake systems can be applied just as well to the lower levels of the added value chain. The 2nd or 3rd tier companies also have to take on a dominant position in their relevant markets in order to maintain their position with the few remaining purchasers at the level above them.

Therefore every supply company should try to get into the top 3 – top 5 in its field, since otherwise independent survival in the market will hardly be possible. The smaller the market is, in which a company is active, the more important is the size of its own market share. In the extreme it is a matter of a niche supplier which must safeguard its market share of 100%.

For an enterprise which does not have such a relevant size there are therefore three possible strategic courses of action:

1. significantly increase its own market share by means of high investment efforts and in its own strength or by taking over competitors;

2. form supplier conglomerates, similar to the model of the OEMs, to enhance its own market power;

3. voluntarily retire from the market or wait until a competitor buys the firm.

A fierce predatory competition is taking place along the entire added value chain, which starts with the earnings pressure of individual OEMs and leads to dwindling margins through to the lowest level of suppliers. Companies which do not want to leave the market, besides choosing the right strategy, will have to meet urgently necessary requirements:

- development of high-quality and above all innovative products,

- safeguard of geographical proximity to the "client" (OEM or system supplier),

- readiness to acquire new business areas in the added value chain,

- formation of development and production cooperations ("clusters") to gain synergy-effects,

- cost cutting, wherever it is possible and makes sense, if possible close to the "benchmark".

Fully in line with the *competitive theory of saturated markets*, "cost cutting" has become the unavoidable destiny of the entire global automotive industry. Naturally cost pressure is highest where both the products are almost identical or at least similar and also the cost levels in the global standard are highest. This is the case in the whole German automotive branch to date. The compelling conclusion is that Germany as a location can only afford the highest cost levels where the products have genuine unique selling positions (USP), and thus are not the same as or comparable to products from low-wage countries.

This applies not only to the final product, the car, but also to all purchased materials and services which are produced by the components supply industry.

6 The future map of global vehicle location – some reflections

6.1 Stocktaking

6.1.1 Industrial states in the maelstrom of globalization

The dimension of economic action has changed dramatically, both for business and politics, within a few decades. Liberal macroeconomists and politicians for centuries focussed above all on describing the advantages of free world trade through the mutual exchange of goods and on helping them to breakthrough in economic theory and in international economic politics through trade agreements (GATT, WTO). Since the 1970's at the latest however, the world economy has been in a new phase of economic division of labour.

In the last two centuries economic politics were almost exclusively about the question of the optimum international *allocation in the distribution of goods and commodities*, that is, about the optimum exchange of goods and the freedom of world trade. The national endowment with natural resources and with the production factors, labour and capital and the industrial production locations in the Old World, was seen as immutable and predetermined. Now that this freedom of world trade has been reached to a great extent by the end of the 20th century, the question arises instead, of the *optimum allocation of locations for the production* of these goods and services. At which locations can identical goods be produced at the lowest cost, that is, with the smallest – measured in money and exchange rate – factor input? This freedom of choice of the best international production location is the *core of globalization*.

The global division of labour which until then had been firmly structured (raw materials from the Third World, industrial products from the First World) gradually began to wobble. More and more developing and newly industrializing countries latched into the international division of labour by means of their own industrial manufacturing operations and developed unnoticcably into compctitors of the established industrial countries. Japan was the pioneer after World War II, and then the Asian tiger states and – within limits – Latin America followed. The catching-up process of the Third World did not gain a new quality until the beginning of the 90's, with the collapse of socialism in central and Eastern Europe and the gradual introduction of free enterprise in China and India. In the meantime over 90% of mankind live with free market principles. That was the beginning of a new chapter of the international economic order, namely the chapter of globalization.

The process of building production sites in low-wage countries from Latin America to Asia is thus in principle no new phenomenon, but has been going on for many decades. In the established industrial countries of the Old World however, this development was hardly noticed by the general public, to say nothing of the highly technical automotive industry with its technological and organizational special-know-how. At least it was not perceived as a threat to its own competitive position, but rather as a more social form of the earlier capitalism: the building of factories in the Third World and thus the creation of employment, income, rising standard of living and increase in the import capacity for expensive goods from the Old World *there*, increase in real income by provision with cheap imported goods without producing increased efficiency or having to accept loss of employment *here*. For many decades the Terms of Trade clearly favoured the highly developed industrial countries, since the developing and newly industrializing countries were not capable of producing high quality industrial goods for their own requirements themselves, let alone emerge as competitors to the established industrial countries in the world market.

This has changed for ever! The age of the "soft" global economic division of labour which favoured the old industrial nations ended abruptly with the collapse of socialism and the bipolar world order at the beginning of the 90's. The globalization of the world economy grew out of the process of internationalization of the world economy with the main players in the big western industrial countries and the scene-shifters in the rest of the (Third) world, when the country with the largest area and the most resources - Russia – and the countries with the largest population – China and India – opened up to the world economy almost over night. Socialism and inefficient planned economies were replaced by economic orders characterized by the free market and founded on competition and private enterprise. This policy quickly bore fruit. These countries now began in their turn to produce industrial goods of world market standard, and to sell them in the established industrial states. For the states and industries of the Old World the age of globalization began.

It is not the appearance of developing and newly industrializing countries on the world stage that is new. It is new that this appearance took place by states with enormous resources and manpower potentials, which had previously isolated themselves voluntarily from the world economy for ideological reasons. It is also new that these NIC's (Newly Industrialized Countries) are no longer countries which passively take over productions from the highly developed countries, to which these have become too strenuous, too expensive, too environmentally damaging, etc. Rather, they are countries which actively raise productions with which they can reckon on having good competitive chance in the world market. The *theory of optimum factor allocation* as a leitmotif for profitable operating was for the first time not restricted to the countries of the northern hemisphere or economic regions of the triad, but embraced the whole world. That is a break in economic history.

It is just these resources and economic potentials of China, India and Eastern Europe / Russia, now steadily but surely pushing their way into the world market, which are being viewed in the established industrial countries with growing fear of what the future will bring. Particularly labour unions and employees, but naturally politicians, too, suddenly see the wealth built up over more than 50 years and above all industrial jobs with the existing high incomes in great danger. *"This new perception works like a shock – virtually over night many realize that in Shanghai there are already more skyscrapers than in New York. And that the developing country of China is today the second most important sales market for a group like Volkswagen."* [74]

Future shocks are not a good basis for mastering challenges. Firstly the question has to be clarified soberly of what has definitely changed for the old industrial states and their people as a result of the globalization process.

Essentially, in this globalized world it is all about the ancient national economic question of the best allocation of scarce production factors. *Which location, that is which country, which region worldwide is best suited for the production of certain products?* Put clearly, this means: At what location can comparable products be manufactured with comparable – or better – quality and at the lowest cost?

---

[74] Ernst&Young (2004a), p. 42.

Within the triad it is the German economy, with the highest industrialization quota of all the western industrial states with outsider competition from previously third world countries, which must deal particularly intensively with the question of *global cost advantages in industrial manufacturing operations.*

For in the course of globalization meanwhile, ever more so-called developing or newly industrializing countries are appearing, in which industrial goods are not – yet- being developed, but are able to be produced with the same quality and productivity as in the industrialized countries themselves, but at considerably lower wage costs. Production in the low-wage countries thus for the producer leads to – usually considerable – relief on the cost side, which in no longer necessarily "paid for" by loss of quality. Thus the "old" locations within the triad have lost a substantial competitive factor, namely the compensation of higher labour costs with a lead in quality and productivity.

The iron proposition of the Theory of Foreign Trade, according to which international division of labour between the First and the Third World in production should ensue corresponding to the respective comparative cost advantages – cloth here, wine there, suddenly gains an unexpected topicality. Only with the small but extremely significant difference that now the NIC's, which previously concentrated exclusively on "wine", but are now going over to producing good quality "cloth" themselves and supplying it to the industrial states with extremely competitive costs. The employees in the "old industrial states", suddenly robbed of their employment privilege of unmatched product quality and production know-how, feel hopelessly defeated in cost competition with the low-wage countries.

And they are, too! Here an example from the automotive components supplies industry. Taking the factor input pattern of an "average supplier" as a basis[75], the proportion of material costs in the total costs is 47%, the proportion of the company's own production wages is 21% (compare Fig. 15 on p. 42), whereby of course a high proportion of labour cost is included in the material costs, too, which must be taken into consideration in the total cost balance sheet. [76] With differences in hourly wages between German wage levels and comparable wages in Poland or China of 35% to 90% the total production costs for suppliers in these regions are therefore

---

[75] VDA (2002c), p. 24 f.

[76] Calculated along the entire automobile added value chain, the wage share in added value amounts to a good 65%.

as far as the figures go – under rigid assumptions – up to about 20-50% lower than in Germany.

Even if this calculation on the basis of the wage component alone considerably exaggerates the actual cost differences and certainly does not correspond with the entire truth of the costs, no supplier may disregard these differences, especially as his attention is increasingly drawn to it by his customers, the automotive manufacturers.

The consequences of this new division of labour in the global economy are well-known: the exodus of industrial production from the Old World into the "promised cost paradise" in Eastern Europe and Asia is in full swing. According to a study by Boston Consulting for BASF[77], currently half of the suppliers in Germany are faced with a shift of production to low-cost countries. thus by 2015 up to 50% of supply parts for automobiles will come from abroad. [78] Eastern Europe is becoming increasingly attractive as a production location, the entire Asian economic area under the leadership of China (and India) is booming and according to economic experts will continue to boom, despite all risk of overheating, with the help of foreign direct investments. [79]

6.1.2 Increasing location attractiveness of the NIC's

The status quo in the global evaluation of production locations for the purposes of the German automotive industry – also with nuances of industry in general – is as follows[80] :

- In global location competition, according to the unanimous opinion of the enterprises, it is not the location with the most favourable strategic position that will win, but rather the location with the most favourable production costs.

- The most important location factors for automobile suppliers are: production costs, labour costs, qualification of the employees, flexibility of the factor work and work attitudes. Remarkable is: however impor-

---

[77] Cited according to Produktion no. 10, 2005.
[78] See also chapter 5.
[79] Roland Berger in a SZ –interview from 2005-01-08.
[80] Ernst&Young (2004a), pp. 3 and 42f.

tant geographic proximity to the OEM is for module and system suppliers, it only plays a minor role for most component suppliers in relation to their purchasers.

Fig. 79. Attractiveness of locations

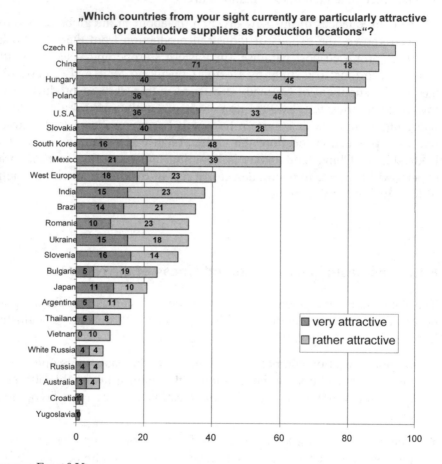

source: Ernst&Young

- German suppliers see China as the world's most attractive location (Fig. 79), well ahead of Eastern Europe (sequence of countries: Czech Republic, Hungary, Slovakia, Poland). labour costs, flexibility and the motivation of the employees, plus the proximity to attractive sales markets – China itself and Western Europe – are the most important arguments.

- The U.S.A. (sales market), South Korea and Mexico (labour costs and sales markets, access to NAFTA) are well in front of Western Europe (and Germany) currently in location attractiveness. Western Europe itself is considered more attractive than Japan, Thailand and Vietnam, it is true, but is a direct competitor of India and Brazil (!).

- Far behind come Russia, White Russia, Croatia and Yugoslavia, countries with an insecure political past. Nevertheless, in the case of Russia, a rethinking has begun among German suppliers, because of the country's favourable development perspective based on its raw materials. The Ukraine, too is meanwhile seen considerably more favourably by foreign investors.

Conclusion: Germany as well as the other triad states currently shows considerable location disadvantages, compared to the low-wage countries, but they (still) have a high industrialization advantage, because of almost two centuries of industrial history. The process of industrial migration has begun, is going on without drama, but is gaining in strength rather than otherwise.

6.2 Perspective of automotive industry in West Europe: the German case

After considering the status quo it is now time to analyse the future perspective of the automotive Industry in general, whether OEM or supplier, in Western Europe and Germany as locations, and with regard to the following aspects:

1. Have Western Europe and Germany had their day as production locations, in the mid and long term?

2. Will automotive industry migrate completely or only partially to East Europe and Asia?

3. Will above all Germany, within Western Europe, be able to survive as an automotive location on the long-term?

6.2.1 The European dimension

Concerning the answers to questions 1 and 2, the IWK has carried out a series of additional, unpublished surveys with OEMs and suppliers of various size and structure, based on the results of the previously cited study by Ernst&Young (2004a). The results of this survey and the public discussion about possible plant closures at DaimlerChrysler (Smart), Opel and Volkswagen have not done anything to change the basic trends already marked out by Ernst&Young, but rather accentuated them considerably.

According to other surveys, half of the European supply industry will have shifted production abroad by 2015. [81] The result is alarming!

Fig. 80. Expectations of the automotive industry, relative to the build-up of new production capacities

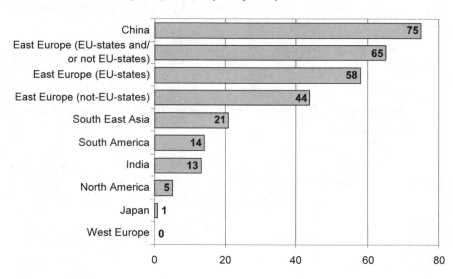

source: Ernst&Young

---

[81] According to Boston Consulting, conf. Produktion no.10 (2005).

According to *Ernst&Young* the Western European high-wage countries have largely had their day as production locations for the automotive industry in the long-term. In the middle-term most of the production locations will be built in China and Eastern Europe: three quarters of those surveyed stated that in the next 10-15 years many new locations will be built in China, 58% mention the Eastern European EU countries, 44% the Eastern European countries not yet in the EU.

Western Europe is not named as a future location by any of those surveyed – „a result that could not be clearer".[82] Reduced to a common denominator, the conclusion of the enterprise survey runs: "Boom in Eastern Europe and China – no chance for Western Europe".

In relation to the automotive manufacturers in Germany this means simply: the BMW plant in Leipzig is likely to have been the last automobile factory to be built in this country. And even that would not have come about, if labour cost levels similar to those in Eastern Europe had not been obtained in internal arrangements with the labour unions. For BMW as a manufacturer of highly priced luxury products these costs were only just economically acceptable. That fits in with statements by Wendelin Wiedeking labour costs are no problem at all for Porsche with its highly priced products[83], while Renault announced at the same time its plans to build a new plant in Russia for new models in the low price segment, and not in France.

One thing is certain: European manufacturers will not undertake any more new investments in the high-wage region of Western Europe and Germany. This applies without saying to Japanese and Korean manufacturers, who want to conquer the European market with cost-effective products despite the fact that it is saturated. And replacement investments in Western Europe are now only made if considerable reductions in labour costs can be arranged with the labour unions. The mobility threshold for suppliers is much lower because of price pressure from their clients and the lack of public consciousness.

Only about 20% of those surveyed by Ernst&Young see the countries of South East Asia as important future production locations, whereby in the case of India the opinion definitely exists that this region could gain increasing significance as a production location for manufacturers and suppliers if the market continues to grow. Accordingly, leading German sup-

---

[82] Ernst&Young (2004a), p. 40.
[83] Acc. Spiegel-Online of 2005-04-05

pliers such as Bosch, Siemens Automotive, WOCO and many others have been active there for a long time already.

It is remarkable that in the expectations / planning of those surveyed Eastern Europe is seen as a completely independent production region and not as an intermediate stop to be replaced by China later. Only 23% of those surveyed believe that China could crowd Eastern European locations out in future.

Fig. 81. Completion or crowding-out

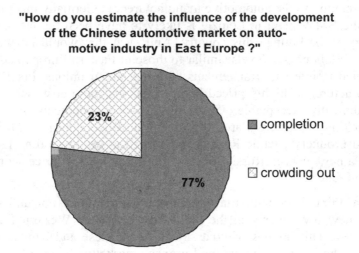

"How do you estimate the influence of the development of the Chinese automotive market on automotive industry in East Europe ?"

23%

77%

completion

crowding out

source: Ernst&Young

This result shows that the automotive suppliers surveyed see Eastern Europe permanently as an extremely strong and attractive location. This allows the implicit conclusion to be made that the suppliers reckon with a distinct increase of OEM production capacities in Eastern Europe. Logistic reasons (just-in-sequence or just-in-time production) require – despite all Chinas labour cost advantages – production sites near to the manufacturers. "*China however, definitely has a distinct advantage as a location for large volume orders that are at the same time labour intensive but with low alteration expenses*".[84] This is already valid today for parts which are

---

[84] Ernst&Young (2004a), p. 40.

exported to Europe and the USA. How much larger will the product range be in the near future, when suppliers are also able to producing for the Chinese automotive industry now in its build-up phase, and will thus be able to produce the necessary batch volumes for profitable investments!

These individual considerations may be summarized in the fundamental question: Which particular segments of the Western European automotive industry (OEMs, 1st tier, 2nd tier, 3rd tier, etc.) will suffer losses in attractiveness and come under pressure to relocate?

The answer is just as simple in terms of economic theory as it is painful in terms of employment politics: there is no future in Europe for all labour and thus wage-intensive activities, such as simple manufacture of components. About 80% of those interviewed shared the opinion that the significance of component manufacture in Western Europe will decrease in future.

In the opinion of about 60% of those asked, final assembly in the supply industry will also migrate to Eastern Europe. This in so far remarkable, that final assembly and quality assurance are inseparably connected. The companies are obviously of the opinion that high quality standards in low-cost countries such as in Eastern Europe and above all China can be achieved and kept even in final assembly. The rest is a question of logistics and transport costs.

Fig. 82. Future areas of competence in West Europe

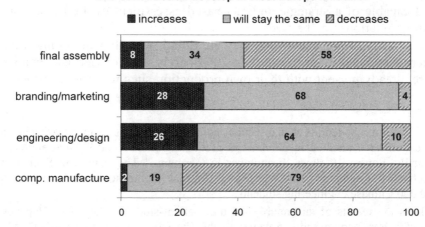

"How will the following core competencies in automobile construction develop in West Europe in future?"

source: Ernst&Young

Besides, complex and capital-intensive supply components are being increasingly manufactured abroad. Parts of research and development are also (by 36% of the companies) being moved to China and Eastern Europe. this does contradict common prejudice, but is plausible in as much as Eastern European universities have for years been training excellent engineers and scientists, who can now be employed at substantially lower wages. Moreover, young Chinese people have for decades been studying increasingly in the West and taking the know-how back with them, just as the Japanese students did two generations earlier. Therefore there is no reason to doubt scenarios in which at the latest when the Chinese automotive industry has reached a "critical" size, research and development from Western Europe, the USA and Japan will be moved to China for cost reasons.

Just this development in China in the last two years shows that the Chinese economy is capable of appropriating technical competence (e.g. in space travel), which was hitherto the domain of highly industrialized countries. Today already, there are 325,000 students of engineering in China, as compared to about 65,000 in the USA and about 37,000 in Germany. [85]

The prospects are more positive for Western Europe regarding other core competence fields in automotive construction: branding / marketing and engineering / design. Only 4%-10% reckon with a migration, while 28% or 26% even believe there will be further immigration into Western Europe. In this case it will probably be mainly by Japanese and Korean manufacturers, which are focussing noticeably on the conquest of the Western European market and are thus building development, design and marketing centres throughout Europe. [86]

It ought to be beyond doubt that the companies surveyed can be considered capable of a strategic and fact-based assessment of the actual competitive relationships between global production locations in the automotive industry. They know what they are doing, why they are doing it and where they need to go. Because in 2004 barely 40% of German suppliers were already present with their own production sites in Eastern Europe or

---

[85] "...the Journal reports that in China, government efforts to improve technical education have resulted in universities graduating more than 300,000 engineers annually, almost 10 times the number in Germany." The American Society of Mechanical Engineers, The Wall Street Journal. Some sources also report of 600.000 Chinese engineering students.

[86] Thus the results of an unpublished multi-client-study of the IWK: "Japanese manufacturers´ conquering strategies in the European automobile market", Munich 2004 (IWK 2004b).

China.[87] Companies such as W.E.T. Plc. (Odelzhausen) have moved their entire production abroad within three years. A further 16% of suppliers are at present planning to relocate.

At present almost one in three of German suppliers is already active in Eastern Europe, one in six in China. 39% of all German suppliers plan to relocate further parts of production to Eastern Europe, 23% - almost one in four – plan similar investments in China.

So what conclusions can be drawn from the results of this survey for the future labour market situation in Western Europe and Germany? Will the tendency be for automotive production to completely pull out of Germany and the other high-wage locations in the West and relocate in the low-wage countries in the East and in Asia?

6.2.2 The German case: location heavyweight of the automotive industry

The spectrum of opinions about the future of the automotive industry in Germany is confusing. It ranges from emotional trembling ("we have no chance") to undeliberated hope (the German automotive industry will stay the world's number one...). In order to find out the truth one must analyse the tested company's coordinates and structure in detail. What is its economic policy structure like, what significance does it have in Germany's economic structure? What dimensions are we dealing with in terms of the potential threat to added value, income and employment?

The automotive industry[88] is indisputably the key industry in the German economy. It is of eminent significance for growth and prosperity of Germany as a business location. It is indispensable as a branch for guaranteeing a high level of income and for safeguarding a high employment volume. But can it continue to play this role in future?

Let us look at the facts before presenting an answer to this question:

---

[87] See also the IWK study: "The German automotive industry in the enlarged EU – engine of integration", VDA (2004b).

[88] Here defined in the delimitation of the Federal Statistical Office as: manufacturer of automobiles and their engines, of tractors, trailers, superstructure parts, vehicle parts and -accessories.

- In the past 6 years more than 5.5 million vehicles were built annually in Germany. More than a third of the entire vehicle production of the European Union is in Germany.

- About 13 million vehicles, or 20% of world automobile production, were built by German automotive companies; the percentage of passenger cars was even 23%.

- In Western Europe in 2004 the new registrations of German company brands was almost 6.8 million cars. This corresponds to a market share of 47%.

- In the Eastern European EU-acceding countries automobile production in 2004 amounted to about 1.5 million units. 59% of this came from German manufacturers. Their share in new registrations of passenger cars in the Eastern European EU-acceding countries amounted to 0.37 million vehicles or about 45%.

- The sales revenue of the German automotive industry more than doubled itself in the 10 years to 2,226 billion euros in 2004. Of the entire sales revenue of German industry in 2004, 18.7% came from the automotive industry. Ten years ago it was only 12.1%.

- About 20% of annual German gross domestic product in the last decade was earned with the product "automobile".

- The number of jobs in the German automotive industry has recovered by 130,000 jobs in the last 10 years after falling from 800,000 to 640,000 at the beginning of the 90's (Fig. 83). During this period the branch has in addition created 160,000 new jobs in Eastern Europe alone. The bleeding in other branches however was more serious, because the proportion of employees in the automotive industry compared to industry as a whole rose in this period from 9.5% to 12.9%. This shows that the branch is still a relatively powerful economic force, despite the problems.

- In addition to those employees directly registered in the automotive statistics, about 1.5 million more are employed in the countless branches upstream and downstream from automotive production (e.g. mechanical engineering, chemicals, etc). Add to those the approximately 3 million employees in the car trade, in the repair business and in the service areas to do with cars, and there are about 5.3 million people in Germany today who make a living, directly or indirectly, from cars, which is still 600,000 more than 10 years ago. In the same period 1.5 million jobs were lost in the other sectors of German industry.

- Thus in 2004 one job in seven in Germany depended on the automotive industry.

- With over 140 billion euros the automotive industry is Germany's strongest export branch. The "foreign trade surplus in cars" was 79 billion euros in 2004. About 80% of the entire German trade surplus (95 billion euros) was thus earned by the automotive industry.

- With a sum of about 100 billion euros 20% of all industry investments in Germany in the last 10 years were in the automotive industry. At the same time an additional 60 billion were invested abroad.

- In the past 10 years the German automotive industry has invested 110 billion euros in research and development. According to surveys by the Endowment Association for the German Economy, an average of over 30% of all R&D expenditure in the German economy is accounted for by the automotive industry.

- The branch underlines its role as a future oriented innovation leader with the leading position in German patent statistics of 3,300 patents, led by Bosch and Siemens.

Fig. 83. Employees in the German automotive industry

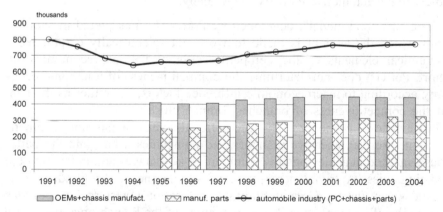

source: VDA, IWK presentation

In short: The German automotive industry is a location heavyweight. Therefore the German Automobile Association was proudly able to give its annual report for 2004 the title "The German Automotive Industry – the engine of the German Economy".

6.3 Determining factors of economic competitiveness

Will the Western European automotive industry be able to remain the engine of the economy? Can Western Europe defend its position, evolved over time, as an industrial location against Eastern Europe or Asia in the 21st century? What will happen in the German Industry?

The answer is not simple. The determination of international competitiveness at the micro-level, that is at the level of the individual company, is relatively unproblematic. It can be demonstrated by the ability of a company to market goods and services successfully, that is with lasting profits, under competition conditions in the international markets. This ability can be seen from the development of balance sheet ratios.

However, it is more difficult to assess the international competitiveness of national economies or branches on the macro-level and to make them operationally comparable and comprehensible. As a rule the structural differences from country to country and from branch to branch are too big to not run the risk of comparing apples with pears and thus coming to highly questionable and vulnerable conclusions. Still less than there is *the* German automotive industry, is there *the* Western European automotive industry. For the sake of simplicity the following observation is restricted solely to the automotive industry in Germany.

The "diamond model", developed by Porter is often quoted in support, with which the attempt is made to work out scientifically certain basic theoretical elements of the competitiveness of national economies. In short, Porter[89] does not determine the competitiveness of a national economy according to macroeconomic influence factors, but rather as the sum of the competitiveness of branches, branch clusters and / or companies. Thus a national economy in the international context is the more competitive, the more competitive branches and companies it has, and the better these succeed in achieving new competitive advantages and in keeping the existing ones. A trivial perception, which no-one can contradict. This would mean that the German automotive location (Western Europe) was internationally competitive if the companies of the branch operating there were competitive. This conclusion is misleading – like the use of the whole Porter model - at the latest when the companies, as for example W.E.T. Plc. (Odelzhausen), only have their headquarters in Germany, but have

---

[89] Porter M. (1999), p. 150f..

already moved all their productive activities to the low-cost countries for the sake of competitiveness.

In the following the attempt is nonetheless made to arrive at approximate statements about the international competitiveness of Germany as an automotive location on the basis of the "Porter model". It is true that it presents a colourful mixture of microeconomic and macroeconomic influence factors without the explanation of genuine causalities between the influencing variables. But still it is suitable for at least making certain basic connections between macroeconomic and microeconomic impact mechanisms schematically comprehensible. For just this interaction is in the end decisive for the competitiveness of national economies. It determines the general conditions for the success and failure of branches and companies in global location comparison – assuming that the qualitative abilities of management personnel in the companies are more or less evenly distributed across all competing companies. How central the role of quality and ethics in management is for the success of a company, even in a difficult market environment such as the global automobile market, is shown more than clearly by Toyota.[90]

The assessment is only about an ordinal and very subjective, not a cardinal an objective evaluation of the current competitiveness of Germany as a location for the automotive industry. The prognosis will be established in the following paragraphs. We have consciously refrained from recommendations to politics and society.

But what does determine the competitiveness of national economies and branches? If one summarizes the essential elements of the competitiveness of the German automotive industry from the previous chapters 1, 2 and 4, one comes to the following abbreviated conclusions for the German automotive industry on the basis of the diamond model (compare Fig. 85 in appendix 4a):

- **Enterprise strategy, structure and competition:**
  - Management in international context definitely cosmopolitan, multicultural and because of strong domestic competition/market structure and the small domestic market traditionally strongly export and competition orientated;

---

[90] Helmut Becker - IWK, Phenomena Toyota – World success by Ethics, unpublished study, Munich 2005

- Considerable progress on the learning curve because of painful experiences with strategic misinvestments within and outside own core business; little tendency to high-risk trial-and-error-investments outside the core business;

- high matured cost sensitivity because of the strong global export competition and the permanent upward revaluation of the D-Mark;

- high tendency to technical innovations as competitive factor;

- strong tendency to rationalization investments because of longstanding labour pressure on costs;

- enterprise strategy orientated to competition and growth, as a rule long-term, not short-term orientated to shareholder value.

On balance: *positive*

- **Factor input conditions:**

  - Good general conditions in infrastructure, legal security, energy supply, bank system etc.;

  - Decreasing quality in education and production of technical knowledge (Pisa study; deficit of scientists/engineers);

  - Factor labour "in the industrial basic load" in international comparison, because of high wage and social welfare expenditure only provisionally competitive;

  - General conditions for utilization of human capital (labour law, tariff provisions, work time settlements etc.) internationally only provisionally competitive.

On balance: *negative*

- **Related and supporting branches:**

  - Dense regional network (cluster) of related and non-related lines of supplier branches;

  - Close spatial cooperation along the entire added value chain between very different branches and enterprises;

  - Permanent technology exchange and information flow between customer and supplier within the cluster allows continuous innovation and improvement process;

- High geographic occupancy with small and medium-sized enterprises and service providers; high industrial density;

- Close cooperation between science and practice; high knowledge transfer.

On balance: *positive*

- **Demand conditions:**

  - Ambitious and competition intensive domestic market is forcing highest quality and technological innovations and top performances;

  - High expenses for research and development as well as compulsion to learning curve effects, force high economies of scales;

  - Small domestic market and compulsion to volume effects force strong export orientation;

  - Progressive saturation of important sales markets.

On balance: *negative*

- **Chance**: *neutral*

- **State and politics:**

  - High regulation and bureaucracy density as consequence of growth and wealth automatism;

  - Creeping annulment of basic principles of free market economy by social overemphasis and structure conservatism;

  - Growing readiness for reforms across politics and society but with lasting disturbance effects by lobbies;

  - Increasing social encrustation by marked corporatism.

On balance: *negative*

If one summarizes these – admittedly very pragmatic and subjective – evaluations, positive and negative location factors are approximately in balance at the moment, but tending towards negative. The competitiveness of Germany as an automotive location has clearly suffered since globalization began, but is probably still among the leaders in the branch average in the global automotive world. Which means:

- The gap to the second best locations has shrunk, both because of home-grown mistakes and by externally determined factors and structure changes.

- From the point of view of many enterprises, the location quality is already beneath the international average. The daily media reports about location shifts and the thinning out of jobs in the *mobile* components supply industry reveal that the substance of the automobile location Germany is already considerably worse for wear.

So much for the stocktaking. It remains unsettled, in which direction the location quality will develop in the short and longer-term. Will Germany be able to safeguard its current *automobile location substance* or will the level continue to decrease distinctly – with all the negative macroeconomic effects for the German labour market?

6.4   Job Engine Automotive industry: Peak Level passed

6.4.1   In the short-term: gradual melting-process

If one weighs up the arguments mentioned above, the following conclusions can be drawn with regard to the threat to the automobile location Germany:

Considering the excellent *quantitative* significance which the automotive branch has in the German national economy as well as in the international structure of the global automotive industry, it seems certain that the German automotive industry *will not forfeit any of its fundamental significance for the German location* in the foreseeable future. The labour cost advantages in Eastern Europe and China can be as big as they like in the next few years, despite all the new investments in these low-wage countries, there is *nowhere* a capacity build-up by OEMs or suppliers in sight, which could *crowd out* anything like even a *negligible part* of the German production and added value volume.

To the delight of economic politics – and the Automobile Association VDA – it may be said: the automotive industry will remain an economic heavyweight in the German industrial structure! There is no way around that!

Nothing can change the fact, either, that the markets in Eastern Europe, China, India or anywhere else in Asia or Latin America will experience rapid growth of their automotive markets and of their automotive industries in the next decades. For the economic development in the old world has shown one thing clearly in the past two centuries: without the automobile as a means of transport and thus without the branch which manufactures these products, wealth and the economic rise of a national economy are not possible. If the above mentioned national economies want to participate in the world's wealth – and with the abolition of socialism and the introduction of free market rules of play for business they have shown that they want it – it is only possible by integration in the global economic division of labour and with higher mobility and more traffic. Without more traffic – no economic development!

Even if it comes to temporary declines in sales and sometimes considerable overcapacities while building up a national automotive industry, as in China in 2004, this is only temporary overheating, as occurred for example in Europe and America while building up the steel industry at the end of the 19th century, too. That blows over. Even so, Production capacities in the dimension of 15-20 million vehicles altogether p.a. will have to be reserved for the automotive markets of this *new world* in the next 20 years. Because the satisfaction of domestic demand will in future take place increasingly from *local production* and not by exports from the triad, especially Germany, as was previously the case, most of the vehicles will come from the build-up of new, local capacities not yet built.

A satisfaction of the growing demand in the form of exports from the high-wage countries of the old world will only be possible within very narrow limits, on the grounds of the given earnings gap, and only for the demand of the upper income levels for luxury brands. In the mass market – the car for Mr. Li, Mr. Sing or Mr. Petrov – only the local production will be able to satisfy the demand, for which further capacities are necessary, which do not as yet exist. But the necessary new investments need time and are in principle no substitute for production here.

Furthermore, the high amount of old stock cars from German production will continue to give German manufacturers guaranteed business from the expected replacement demand in the most important foreign markets and domestically. Customer loyalty – understood here as *resale rate* – is still high, with almost all manufacturers, even if the level varies from manufacturer to manufacturer. But it is noticeable that it is decreasing for all European manufacturers. According to psychologists and market experts, in the course of the advancing market penetration of foreign – above all Japanese and Korean – brands with increasing rationality, that is decreasing emotionalization of the purchasing decision, the customer is increasingly becoming a "*traitor to the fatherland*", always on the look-out for the best buy. As we said: in spite of everything, the customer bond is currently still so strong that the German automotive industry can reckon with an annual replacement demand of between 2.5 and 2.8 million vehicles in the trend of the next few years in the domestic market alone. Even without growth a safe "employment bank".

This demand cannot be even nearly met in the next few years by the automotive plants in Eastern Europe and China which exist today. However it is unmistakeable that the future domestic replacement demand will be increasingly met by low-price offers from the foreign plants of German and French, but also Japanese and Korean manufacturers in Romania, Poland, the Czech Republic, Hungary, Russia, Brazil, and Mexico. This will then be at the expense of domestic and Western European employment. Opel seems to be making a start.

In the short term there may be the one or other manufacturer in Germany - as a whole or just with certain model series – which has structural location problems, but the location Germany as a whole is not endangered.

However, from the German point of view this does not mean that one is allowed to rest on one's laurels. For the competition from Japan and Korea – previously only carried forward by high reliability and favourable price-performance ratios at medium innovation level – is advancing due to considerable efforts in design and technical innovations (e.g. hybrid propulsion) solidly into the core competence fields of the German manufacturers. In the middle and long term this gives rise to a distinct potential threat, which is however not specific to a particular location, but depends on the better offers of the competitors. But the branch has already fully recognized this challenge and there is no reason for location pessimism on the grounds of a loss of technological competitiveness.

Basically it may be assumed that the innovative strength of the German automotive industry as a whole does not need to fear a sudden loss of customer favour either domestically or internationally.

There are many reasons for the innovative strength of the German automotive industry. One of the essential reasons which is often overlooked in the relevant analyses, is the dense network of relationships between OEMs, suppliers of completely differing degrees of integration and the other intermediate performers, who are "hidden" in other branches such as synthetics production, mechanical engineering, the IT industry, or the consulting business, but all make a living from the automobile. In short: It is the industrial *clusters*[91] of the German automotive industry which have given the branch such a strong global competitive position right up to today.

The center of the cluster is always one or more plants from different OEMs, as for example in the automobile triangle *Dingolfing / Landshut – Regensburg – Ingolstadt,* around which the supply companies settle. Thus it does not astonish that according to an analysis of the iwd (Köln) that the following towns / regions[92] have the stated proportions of employees subject to social insurance contributions in the automotive industry:

- Ingolstadt (Audi) 22.2%

- Landshut /Dingolfing (BMW) 20.7%

- Braunschweig (Volkswagen) 19.9%

- Stuttgart (DaimlerChrysler) 10.3%

The enterprises and jobs which are indirectly connected to the branch must be added to these, for example the baker who delivers the bread rolls for the factory canteen.

About 5.500 direct components supply companies of various kinds are available in Germany for creating networks. The important thing is that they work together in networks, although they appear in the market as competitors.

An automotive cluster becomes almost unassailable if training centers such as technical universities and advanced technical colleges as well as research institutes and facilities with a close connection to the relevant industry join the cluster. That is usually the case in Germany. It is just this close cooperation between scientific research and practical implementation that has been the strength of the German automotive industry so far. As practised very successfully by the Munich TU or the Ika-Aachen this cooperation allows a continuous flow of technical progress and innovations.

Innovations, which as a rule are today made by the suppliers as product specialists, almost always arise in trial-and-error procedures. That is, their practical application and feasibility must be worked out in close cooperation with the OEMs as purchasers. The development departments of the

---

[91] **Clusters** generally mark a unit considered as a whole, consisting of various details. In the automotive industry a cluster is a spatial concentration of smaller and bigger enterprises of the branch, whose core is always an OEM.

[92] See iwd (2004a), p. 4f.

suppliers very often work in the same premises as the development engineers of the OEMs.

It was only because of this close spatial cooperation that the German automotive industry was able to exhibit a higher rate of innovation and higher quality than the foreign competition in the last 10 years, and thus was able to maintain and extend its global market position, despite the cost disadvantages in Germany.

Nevertheless, there is a danger that this advantage at some time or other could turn into a disadvantage, if the marginal cost of innovative output for the customers becomes greater than its marginal utility.

Conclusion:

A lasting threat to Germany as production location can be ruled out in the short-term. The bonding power of the German automotive cluster is too strong to be forced out by differences in labour costs.

But the German automotive industry has definitely passed its production and employment peak. Capacity cuts and the drastic reduction of employment by German automotive manufacturers which are rich in tradition (e.g. Opel), and a gradual relocation of production or the building of new capacities only in the new Eastern European EU acceding countries or the other parts of Eastern Europe (e.g. Continental in Romania) show that the supply industry is partly leaving the German location already.

It is unmistakeable that the *"caravan of German suppliers"* is wandering further to the East or China day by day. It is however so far only a caravan and not the whole *"oasis"*.

Put clearly: the wage-intensive non-cluster-bound manufacturing operations are all migrating little by little to low-wage locations for competitive and cost reasons – or are already there. This is clearly at the expense of employment in Germany. All supply companies which produce wage-intensive individual parts and are not necessarily an integrated part of an added value cluster will relocate production to Eastern Europe and / or China – as far as they have not done so already.

System suppliers working in immediate manufacturing association with the OEMs will stay here. Module and individual parts suppliers migrate, as far as aspects of delivery safety and transport costs do not stand in the way (see chapter, Consequences for the Components Supply Industry p.163ff.). If suppliers are too small and financially weak to deal with a relocation of production they will be taken over or will withdraw completely from the market.

According to rough estimates by the IWK, about 150,000 of the 780,000 jobs in the German automotive industry (OEM and supply industries) will be lost in the next 10 years. Of the approximately 380,000 jobs in the supply industry almost one in three could be cut due to relocation abroad.

This sounds spectacular, but it isn't – apart from the fact that branch experts find this estimate rather too conservative. On the one hand because the transfer of jobs to a new location has many aspects. As a rule it takes place gradually and only from the "edges" of the branch, with the marginal suppliers, which are anyhow not in the public eye.

This is of course in the interest of the enterprises concerned, themselves – simply because of the negative effect on their public image and for bypassing resistance from labour unions. The public debate about the increasing foreign investments of German enterprises, these "traitors to the fatherland" and their "unrestrained greed of gain" has meanwhile erupted completely. If it is then said that "the economy is there for the people, and not the people for the economy", then it is completely underestimated that this ethical principle has not changed a single jot. Just that the economy is now there for the people in Eastern Europe and China, and no longer for those in the high-wage location of Western Europe, who want to be supplied with cheap products.

On the other hand over 600,000 employees still remain active in the branch directly in Germany, according to the calculation above – not really a negligible quantity!

The automotive industry will thus remain the most important branch of the "old economy" in Germany for the foreseeable future.

6.4.2 In the long-term: reduction to essentials

As shown in the previous chapters the question of whether Germany and Western Europe can remain long-term locations for automotive production is substantially dependent on whether the OEMs can hold their sway as the central core of the automotive clusters in this economic region.

It will only become critical for Germany as an automotive location when the actual core of automotive production – assembly and development – pull out of Germany and other Western European high-wage countries. That is to say, if the OEMs were to successively thin out their German plants and in the end shut them down completely. This would then automatically force the larger 1st tier suppliers to migrate. As long as the OEMs (still) stay at the location Germany with essential "real world" functions – production and development, so long will a considerable part of the supply industry also stay in Germany.

The central point is therefore the competition and earnings situation of the OEMs themselves. They are the Achilles' heel in the stability of the production location Germany, not the supply industry.

The central question is therefore: what are the critical points for the location steadfastness of the German automotive manufacturers?

The results of our examinations (chapter 1-3) have shown that in future more and more cars will be produced in Eastern Europe and China and thus the global market share of these regions will continually increase. In Eastern Europe, led by Poland, the Czech Republic and Hungary, the first clusters in the pattern described above are already well advanced. [93] The supply of the Western European sales market from this region, above all by Japanese and Korean, but also by French manufacturers, is about to begin, and the supply by German manufacturers themselves, such as Opel and Volkswagen (Skoda) is well under way and very successful. The formation of its own automotive clusters has already started in China, too. The export plans of Chinese automotive manufacturers (e.g. Brilliance, SAIC et. al.) for vehicles of purely Chinese manufacture are becoming known – today still smiled at by the established manufacturers, but that was the same when the Japanese and Korean automotive industries first started up around 50 and 30 years ago.

All the German manufacturers together are under considerable competition and cost pressure. The massive cost reduction and restructuring programs first pushed through by all German manufacturers in 2004 - publicly and spectacularly for some (DaimlerChrysler, Opel and Volkswagen), for the others below the surface and closed to the public (Audi, BMW, Ford, and Porsche) – show a high level of insight and considerable need for action to safeguard international competitiveness or even sheer existence. Or as the great German poet put it: *"Obeying necessity and not one's own desire"*.

---

[93] VDA (2004b).

Since as early as 1998 domestic passenger car production has no longer been growing, but fell by 150,000 units to 5.2 million in 2004. If exports had not risen in the same period by 400,000 units from 3.3 million to 3.7 million the decline in production would have been considerably greater.

The globalization of the German automotive industry, recognizable in the construction of production sites abroad, is advancing further. Even the Association of the German Automobile Industry has to concede "... that *the globalization process ... is irreversible, if by this the international networking of production and procurement is meant, and it is increasingly revealing the weak points of handed-down structures. In the last few years the location factors determined by costs have damaged domestic competitiveness and thus contributed to the relocation of production in mort attractive regions."* [94]

As a result the internal German added value share per vehicle from domestic manufacture has been continuously decreasing. From 72 % at the beginning of the 90's to meanwhile 60%.

Location arbitrage and strategic capacity investments for opening up new markets have led to the German automotive industry now producing in 45 countries. The foreign production of German OEMs rose by a third between 1998 (3.3 million units) and 2004 (4.4 million units), while at the same time domestic production shrank. About 45% of all cars of German brands are already manufactured abroad today.

The most important reason for German foreign manufacturing is – as shown above – the high domestic pool of labour costs. Here, just as with the shortness of working hours, the German automotive industry holds an international top position, as is well known.

At 33 euros per man-hour the automotive labour costs in Germany in 2003 were 20% higher than in the USA and Japan, 50% higher than the average of Western European competitors and 550% higher than comparable costs in the new EU countries – to say nothing of newly industrializing countries in Asia, such as China.

Moreover, in the German automotive industry the annual labour time is at 1,430 hours 300 to 500 man-hours less than in other countries. These cost disadvantages cannot be balanced out by higher productivity, even if

---

[94] VDA (2004a), p. 13.

the sales revenue per employment hour rose by almost 70% in the German automotive industry in the past 10 years. [95]

It follows from this that the global cost and profit pressure on the German OEMs has obviously become so strong that *cost cutting* merely in the "mobile part of the added value chain", namely the suppliers, by relocation in low-wage countries is no longer enough. The OEMs must now optimize their own cost structures. *Location arbitrage,* that is the construction of production capacities in low-wage countries, is a last resort here. For it is absolutely clear that as far as labour costs are concerned, a high-wage country such as Germany can have no chance, structurally, of competing seriously against, for example, China.

Therefore the problem of labour cost arbitrage gains a new dimension in the meantime. Other than in the past, production capacities have been developing in the low-wage countries in the last 10 years, which are *no longer only for domestic demand* (e.g. Eastern Europe and China), *but also – or even solely – for exports to the volume markets of the old industrial countries* (e.g. from Romania, Brazil). While the traditional manufacturers were previously able to divide the highly developed volume markets of the triad more or less between themselves, for the first time *genuine outsider competition from the former developing and newly industrialising countries* poses a threat. Purely related to costs, the old production locations cannot keep step here, other competition parameters must take effect if they want to stay "in the game".

Thus all the added value chains in the traditional automotive countries experience a new quality of pressure to adjust. For the first time the OEMs themselves are under pressure – whether luxury or mass producers.

Completely independently of the risk potential from this newcomer competition from the low-wage countries, a second factor is the additional threat from the advancing oligopolistic competition of the established manufacturers in the triad. In the automotive markets of the old world with strong signs of saturation and the OEMs' growing overcapacities this cut-throat-competition is becoming increasingly fiercer and is meanwhile making itself felt in Germany.

Massive discount offers, which were confined to the US market after 2001, have reached Europe and the German automotive market in the meantime. The discount battle for *homo automobilis* has definitely taken hold, even on the territory of *homo automobilis nobilis.*

---

[95] VDA (2004a), p. 14.

Distinctly falling profit margins in the core business of all the OEMs, even the so-called noble brands, have partly led to spectacular cost reduction programs with massive workforce cutbacks. In the volume segment some traditional automotive manufacturers have already been making high losses, sometimes even increasing, for years in operative business and are slowly but surely moving towards withdrawal from the market, at least if the parent companies, which have their own problems, no longer want to keep up the cross subsidization.

The pressure on margins which the OEMs are faced with is passed straight on to their suppliers, or to put it a better way, they try to pass it on. the price pressure at the 1st tier level is passed on like a cascade to the downstream supply levels (=cascade competition), which finally leads to the withdrawal (takeover or bankruptcy) of marginal suppliers from the market.

The product specialists, innovation pioneers and niche suppliers will be able to avoid the cost pressure from the OEMs the longest, but only for so long as their customers value the benefits of cost-intensive product innovations more than the price to be paid for them. The limits of the customers' capacity are moving nearer, in times of stagnating or even falling real income, innovations with higher price margins – and list prices – are alone no longer enough to save the margins. The OEMs must deal with their own costs – they are not used to that and are thus finding it extremely difficult.

Thus it is the location shifts on the supply side which are still the focus point for the foreseeable future. When the first European manufacturer - the end of Rover is a special case – falls victim to the merciless crowding-out competition and moves its production to Eastern Europe cannot be forecast.

As expected, the relocation of part or all of production to the so-called low-wage countries – here especially Eastern Europe and China - has proved particularly cost-efficient for the suppliers. Both regions are given excellent *location marks* by German suppliers (Fig. 84).

This is not new. Well-known components supply companies (e.g. Bosch, Continental, WOCO, Webasto, etc.) have sufficient experience of the cost advantages of foreign production locations at invariably high quality, due to having built their own factories in the wake of the their traditional OEM customers setting up new production sites for CKD and CBU assembly. Thus positive experience exists, and insurmountable entry barriers to the low-wage locations no longer exist, if one ignores investment hindrances specific to the country, such as corruption, bureaucracy, etc.

Fig. 84. Evaluation of the locations Eastern Europe / China

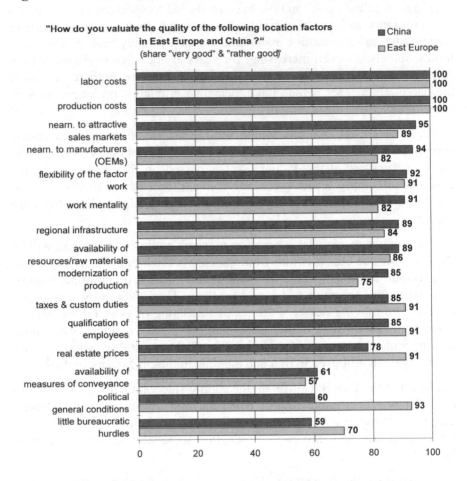

source: Ernst&Young

The fact that German automotive manufacturers construct plants for the strategic development of markets abroad is not new, as shown by the examples of BMW in South Africa (1973) and the USA (80's), or Daimler Benz in the USA (80's), Volkswagen in Brazil and Mexico (60's and 70's) etc. Numerous regular suppliers followed their OEM customers abroad. But it was never a matter of cost-driven production shifts by location substitution, but rather of growth-driven production expansion by location development. Thus the Bosch group, already in China since 1990, built a first joint venture into a holding with 11 subsidiaries, 9 joint ventures, 21 production sites and 6 distribution companies with altogether 13,000 employees.

All the German manufacturers and the majority of the suppliers therefore already have experience in the building and operating of foreign automotive plants. It is only the motives for the foreign investments that have changed: while they used to be mainly active or passive market development investments, today they are frequently cost cutting investments, above all in Eastern Europe. In China however, the market development motive is probably currently the main emphasis. From the point of view of the German location it does however present a problem if big suppliers such as Bosch, the Schaeffler group (INA, FAG, LuK) or the Mahle Company develop not only production but also development activities in China.

Thus there are no inhibitions from a business administrative point of view as to increasing this investment relocation process which is now under the sign of the margin squeeze. That exactly is the long-term threat for Germany as a location! If the OEMs no longer consider the profit margins aimed at lastingly achievable in the German location – after weighing up all the pros and cons of relocation abroad – they will migrate. Not from greed of gain, but simply for self-preservation! That is where the world of the most immobile real capital differs a lot from the world of the most mobile financial capital. Whereby some economists would consider it very useful, if the latter would sometimes make the former get a move on.

Nevertheless, there are also delaying factors! As much as Germany is threatened as a production location in the long-term for the above reasons, there are still many positive factors:

- The problem-consciousness of the threat now exists for all those responsible for the location in politics and economy. It is just the cost reduction and restructuring programs taking place throughout all companies in the branch which are the proof that the quality of the location is being worked on. And this is happening with the growing consent of the workforces and company councils concerned, which were previously always considered a slowing factor for sensible measures. The more quickly these programs take hold, the better the location quality for the whole branch will improve compared to foreign locations and competitors.

- Measures to improve the quality of the German location are taking place everywhere, for example to increase efficiency by higher production, longer working hours at unchanged income, absolute wage cost cuts, etc. The insight of the people concerned is growing everywhere that their jobs can be saved by their own contributions. In addition, there is the high level of qualification of German employees, which is certainly something to be proud of in the international competition.

- In scope and depth Germany disposes of a historically grown and above all scientifically founded and secured cluster quality, which can hardly be developed in this form in low-wage countries within a few decades? Fraunhofer –Society, Max-Planck-Institutes, university establishments such as the Ika at the RWTH Aachen or the Munich TU, to mention but a few, form a technical medium whose simple existence and operative capability maintain the competitiveness of the German automotive industry. The automotive industry was, is and remains in future a high-tech industry. That makes a fundamental migration of the automotive industry from Germany almost impossible. But whether in 20 to 30 years China or India will have a similar scientific-technological infrastructure to Germany today is difficult to forecast. Everything is possible.

6.5 Summary

Even if the truth is painful: at the location Western Europe, headed by Germany, the automotive industry has passed its peak. It will remain the parade-horse of the German economy for the foreseeable future, but *the growth engine has already lost speed distinctly.*

So far it has been above all the German automotive suppliers which as the pioneers of globalization have been leading among all German enterprises in the process of the internationalization of production and employment, and are the most advanced in this point. Within the foreseeable future more than half the German suppliers will be present with their own production sites in Eastern Europe and China, some even with complete relocation. The reasons for this have been explained in the preceding chapters.

As it seems, the suppliers have in the meantime already very much exhausted their possibilities as *cost buffers for the OEMs at the German location.* The compulsion to cost-induced location adjustment is now reaching the OEMs themselves in Western Europe and Germany. The measures taken by all the German manufacturers to reduce and adapt capacities and increasingly to build up new capacities in cheaper foreign countries are an unmistakeable sign. The latest announcements by Renault and Volkswagen about the construction of new vehicle factories in Russia are indisputable proof.

Market adjustments are inevitable and they have always happened. And hand-in-hand with them is the *location adjustments*. Painful capacity adjustments by means of plant closures, staff reductions and restructuring measures in the whole added value chain have already been started in the German automotive industry or are imminent, according to our information.

All that will indeed weaken employment in the automotive location Western Europe and Germany, but will strengthen it qualitatively, namely due to concentration on a higher added value. A lasting existential threat to the automotive industry is thus not exogenously presaged. Whereby however the global entry of an independent Chinese automotive industry coming up on the horizon represents the large unknown quantity in the forecast game. If one takes this competition only partly seriously, then there is nothing which speaks for an improvement, but much that speaks for things becoming worse. So the all-clear cannot be sounded!

Appendix

Table 27. Top-100 automobile suppliers

Rang	Unternehmen	Automotive Um-sätze 2003/2004 in Millionen USD[1]	Rang	Unternehmen
1	Robert Bosch	31.400	51	Pioneer
2	Delphi	28.700	52	Alps Automotive
3	Denso	23.172	53	Koyo
4	Johnson Controls	20 500	54	Saint-Gobain
5	Bridgestone	18 280	55	Tokai Rika
6	Michelin	17 347	56	Toyoda Gosei
7	Visteon	17 097	57	Karmann
8	Lear	15 746	58	Cummins
9	Magna	15 345	59	Dura Automotive Systems
10	Aisin Seiki	14 501	60	Pilkington
11	Goodyear	13 362	61	Brose
12	Continental	12.426	62	Getrag
13	ThyssenKrupp	12 359	63	Rheinmetall
14	Siemens (VDO + Osram)	11.815	64	AsahiGlass
15	Faurecia	11.540	65	NTN
16	TRW Automotive	11.308	66	Showa
17	Valeo	10 527	67	TI Automotive
18	ZF Group	8 346	68	Hayes Lemmerz
19	Dana	7.918	69	Hutchinson
20	ArvinMeritor	7 788	70	Freudenberg
21	Yazaki	6 375	71	Key Automotive
22	DuPont	6 087	72	Draxlmaier
23	Federal Mogul	5 546	73	SKF Automotive
24	Autoliv	5 301	74	Illinois Tool Works
25	Matsushita Electric	5 016	75	PiasticOmmum (Inergy)
26	Motorola	4 870	76	Stanley Electric Group
27	Calsonic Kansei	4.771	77	Webasto
28	GKN	4 706	78	3M Automotive
29	BASF	4 564	79	Degussa
30	Schaeffler	4 389	80	Metaldyne
01	DPG Industries	4.290	81	Eberspächer
32	Collins&i Aikman	0 001	82	NewVenture Gear
33	Mitsubishi Electric	3 774	83	Tachi-S
34	Hitachi	3 773	84	Rieter
35	Tenneco Automotive	3 766	85	Flex-N-Gate
36	Mahle	3 701	86	Tomkins
37	American Axle & Manufacturing	3 683	87	Alcan
38	Magneti Marelli	3.655	88	NSK
39	Honeywelf	3 650	89	Edscha
40	Hella	3 525	90	Textron
41	Cooper Tire & Rubber	3 514	91	Kostal
42	Pirelli	3 386	92	Dow
43	Takata	3 241	93	Trelleborg
44	Behr	3.153	94	Schefenacker
45	Benteler	3.111	95	Mann + Hummel
46	BorgWarner	3 069	96	Meridian Automotive
47	Eaton	2.962	97	Oxford Automotive
48	Bayer	2 820	98	Mitsui Mining & Smelting (Gecom)
49	Tower Automotive	2 816	99	F.Tech
50	Alcoa	2 800	100	Teksid Aluminum

sources: enterprise statements, business reports, AP official searches,1 turnover for business year 2003 or 2003/2004, automobile production (February 2005).

Appendix 2

Table 28. Automobile manufacturers in international comparison

group	country	sales revenue	yield [3]	market expectation [4]	rank (acc. to sales)
General Motors	U.S.A.	185.524.0	5.52	8.15	1
Ford	U.S.A.	164.196.0	2.75	5.50	2
Toyota	JPN	151.013.8	7.54	3.46	3
DaimlerChrysler	DEU	147.531.2	1.39	3.72	4
Volkswagen	DEU	92.486.5	2.07	0.68	5
Honda	JPN	74.445.7	8.92	3.27	6
Nissan	JPN	64.419.1	14.50	4.42	7
Peugeot	FRA	63.501.5	4.97	0.36	8
Fiat	ITA	54.315.5	-1.97	3.32	9
BMW	DEU	47.866.9	5.39	4.25	10
Renault	FRA	42.618.8	1.18	-4.81	11
Hyundai	KOR	38.297.9	9.16	7.62	12
Mitsubishi	JPN	35.431.4	5.77	1.27	13
Mazda	JPN	26.596.1	-1.88	3.68	14
Suzuki	JPN	20.055.5	5.88	2.93	15
Fuji Heavy Industries	JPN	13.128.3	3.52	2.40	16
Kia	KOR	13.049.9	8.42	7.00	17
Isuzu	JPN	13.045.2	-0.47	4.69	18
Yamaha	JPN	9.305.2	5.88	2.92	19
Daihatsu	JPN	9.062.1	-0.36	2.69	20
Porsche	DEU	6.728.3	14.68	10.97	21
Harley-Davidson	U.S.A.	4.624.3	20.10	11.89	22
Astra International	IDN	3.267.8	10.08	10.71	23
Tata Motors	IND	3.001.6	10.25	10.47	24
Proton	MYS	2.438.9	15.59	-0.89	25
China Motor	TWN	2.365.7	14.87	8.94	26
Chongqing Changan	CHN	1.695.6	19.47	6.45	27
Maruti Udyog	IND	1.594.5	2.33	9.58	28
Thor	U.S.A.	1.571.4	23.68	11.73	29
Yulon	TWN	1.406.5	4.41	8.38	30
Brilliance China Automobile	HKG	1.221.5	11.65	6.29	31

Hero Honda	IND	1.121.5	35.72	18.37	32
Mahindra & Mahindra	IND	1.104.9	5.73	9.98	33
Faw Car	CHN	1.082.0	7.25	8.40	34
Bajaj Auto	IND	922.9	9.18	8.87	35
Winnebago Industries	U.S.A.	845.2	13.85	12.23	36
Denway Motors	HKG	196.8	31.56	13.03	37
FFP	FRA	6.7	-2.21	NA	38

source: Manager Magazin (2004-23-07); data base: only listed enterprises, sales and yield: status quo 2003; 2) sales 2003 in million US-Dollar; 3) current Cash Flow Return on Investment (CFROI); 4) five-year-forecast for the CFROI on base of current share prices.

Appendix 3

Table 29. Global production 2004 and 2009 by manufacturers

design parent	2004	2009	CTG[1]
General Motors	8.894.678	9.102.367	1.96%
Ford	7.682.638	8.056.006	3.51%
Toyota	7.596.181	9.561.181	18.50%
Renault/Nissan	5.756.286	6.888.894	10.66%
Volkswagen	5.172.769	6.411.397	11.66%
DaimlerChrysler	4.108.749	4.592.489	4.55%
PSA	3.509.105	3.894.586	3.63%
Honda	3.152.970	4.192.852	9.79%
Hyundai	3.050.715	4.916.968	17.57%
Suzuki	2.534.195	2.988.799	4.28%
Mitsubishi	1.808.397	1.770.574	-0.36%
Fiat	1.744.656	2.667.662	8.69%
BMW	1.232.730	1.485.666	2.38%
AO AvtoVAZ	852.845	787.923	-0.61%
total (difference)	*59.312.687*	*69.935.868*	*(10.623.181)*

source: Automobilproduktion; [1]CTG = Contribution to Growth; according to CSM-definition "share in overall growth of the production output of the world-wide automotive industry from 2004 until 2009."

Appendix 4a

Fig. 85. Diamond approach according to Porter

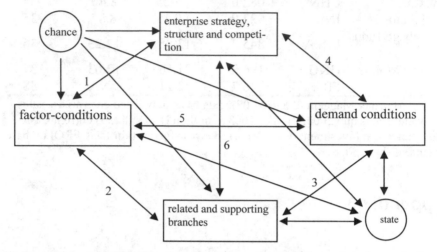

(1) Domestic competition has the effect that enterprises themselves invest in factor creation and maintain relationships to institutions, in order not to be left behind. Workers tend to train branch specifically, if several employers exist. People from factor creating institutions such as universities and laboratories often set up enterprises themselves.

(2) Related and supporting branches partly need the same general production factors, knowledge and infrastructures and invest in factor creation. Qualifications, special knowledge and technology, which develop in a branch, are also useful for related and supporting branches.

(3) Internationally active suppliers and related branches guide global demand to the branch. Size and growth of domestic demand may increase the width and specialization of supporting branches.

(4) Several offensive competitors make the purchasers more choosy and more ambitious. Competing enterprises invest in sales and expand demand. Saturation ensues earlier, which leads to innovations and internationalization attempts. A branch in which demand is high attracts many entrepreneurs and competition increases.

(5) Disproportionately strong demand causes social and private investments in factor creation. Furthermore the state is more prepared to support factor creation. World leading training centers attract foreign students, and demand rises.

(6) Internationally successful businesses guide global demand to the suppliers. Enterprises set up their own components supply businesses. Conversely, suppliers often enter the downstream branches , too. They bring in information, knowledge and new ideas. Related branches diversify with increasing saturation of the basic branch and enter the branch.

Appendix 4b

Fig. 86. 5-forces-model according to Porter

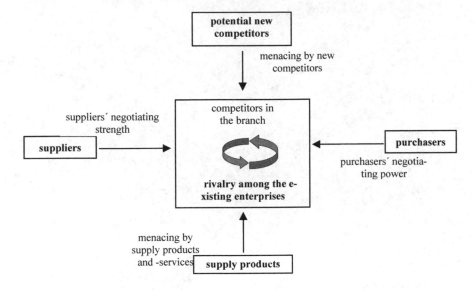

List of figures

List of tables

List of abbreviations

ABS	Antilock braking system
ACC	Associate candidate countries
ACEA	Association des Constructeurs Européens d' Automobiles (European Automobile Manufacturers´ Association)
ASEAN	Association of South East Asian Nations
BAIKA	Bavarian Innovative- and Cooperation Initiative for the Automobile Components Supplying Industry
BCG	Boston Consulting Group
BDI	Bundesverband der Deutschen Industrie (Federal Association of German Industry)
BRIC	States Brazil, Russia, India and China
CAR	Center Automobile Research
CBU	Customer Business Unit (final assembly)
CES	Current Economic Situation
CFROI	Cash Flow Return on Investment
CKD	Completely Knocked Down (final assembly)
EBIT	Earnings before Interests and Tax
EBITDA	Earnings before interest, taxes, depreciation and amortization
ESC	Electronic Stability Control
ESP	Electronic Stabilization Program
FAZ	Frankfurter Allgemeine Zeitung (Frankfort General Newspaper)
FAST	Future Automotive industry Structure
FERI	Finance and Economic Research International
FIZ	Research and Engineering Center
FTD	Financial Times Germany
GATT	General Agreement on Tariffs and Trade
GDP	Gross Domestic Product
IFA	Institute For Automobile Economy
IKA	Institute for Motor Transport Service of the RWTH Aachen
IKB	German Industry Bank

iwd	Information Service of the Institute of German Economy, Cologne
IWK	Institute for Economical Analysis and Communication, Munich
ISI	IWK Survival Index
IT	Information Technology
JAMA	Japan Automobile Manufacturers Association
KBA	Automobile Federal Office
LMU	Ludwig-Maximilians-University Munich
MOE	Middle and East Europe
NAFTA	North American Free Trade Area
NIC	Newly Industrialized Countries
OICA	Organization Internationale des Constructeurs d'Automobiles (International Organization of Vehicle Manufacturers)
OECD	Organization for Economic Cooperation and Development
OEM	Original Equipment Manufacturer
OPEC	Organization of the Petroleum Exporting Countries
PSA	Peugeot-Citroën-Group
PWC	PricewaterhouseCoopers
R&D	Research and Development
ROCE	Return on Capital Employed
RWTH	Rhine-Westphalian Technical University Aachen
S&P	Standard and Poor's
SAIC	Shanghai Automotive industry Corporation
SUV	Sport Utility Vehicle
SZ	Süddeutsche Zeitung (South German Newspaper)
TOT	Terms of Trade, trade relationship of domestic and foreign goods
TPS	Toyota Production System
TU	Technical University
VDA	Verband der Automobilindustrie (Association of Automotive industry)
WestLB	Westdeutsche Landesbank (West German Regional Bank)
WTO	World Trade Organization
YoY	Year over year, annual rate of change
ZEW	Center for European Economical Research

Literature

ACEA (2005): *New Motor Vehicle Registrations*. URL: http://www.acea.be/ASB/Download.nsf/ CategorizedView-Files?OpenForm&Language=English&cat1=6

ADAC (2004): *Der ADAC-AutoMarxX im Dezember 2004 – Die Ergebnis – Übersicht*. URL: http://www.adac.de/mitgliedschaft_leistungen/motorwelt/automarxx/gtamd2004/default.asp?ComponentID=101758&SourcePageID=101788%230

ADAC (2005): *ADAC-Preis „Gelber Engel" 2005*. In: ADACmotorwelt, Heft 2/05, Februar 2005, S. 12-23.

ARD (2004-10-28): *Autohersteller in der Kostenfalle*. URL: http://boerse.ard.de/druck.jsp?key=dokument_69384

AUDI (2004): *Auto & Konjunktur*. Materialien zum Vortrag des Oestricher Kreises vom 21 bis 22. Oktober 2004 in Budapest.

Automobil-Produktion (2004a): *News – Toyota stärkt FuE in Europa*. in Automobil-Produktion, Ausgabe 4, April 2004

Automobil-Produktion (2004b): *Osteuropäisches Roulette*. in Automobil-Produktion, Ausgabe 6, Juni 2004

Automobil-Produktion (2005): *Im Fokus: DaimlerChrysler*. In: Automobil-Produktion, Ausgabe 2, Februar 2005

Automobilwoche (2005-01-17): *Große Herausforderung" BMW-Finanzvorstand zu den diesjährigen Chancen und Risiken*. In: Automobilwoche Nr. 1/2 17.01.05, S. 13.

Automobilwoche edition (2004a): *Die Grenzen der Nischen – Nischen ohne Grenzen?* In: Automobilwoche edition, November 2004, S. 74.

Automobilwoche edition (2004b): *Es lebe die Nische. Markt Segmentanalyse*. In: Automobilwoche edition, November 2004, S. 8-10.

BAIKA (2003): *Technologietrends in der Automobilindustrie und ihre Auswirkungen auf die Zulieferindustrie*. Überblickspräsentation auf Einladung der „Regionalmanagement Wirtschaftsregion Bamberg-Forchheim GmbH am 29. Oktober 2003, Schlüsselfeld.

Bayern Innovativ (2002): *Technologie-Trends in der Automobilindustrie.* Bayern Innovativ Gesellschaft für Innovation und Wissenstransfer, Nürnberg.

BCG (2004): *Produktionsstandort Deutschland – quo vadis? Fertigungsverlagerungen – warum es sie gibt, wie sie sich entwickeln werden und was wir dagegen tun können.* The Boston Consulting Group, München.

Becker H. (2001): *Logistik – Ein Überblick*, URL: http://home.t-online.de/home/becker2/log2_1_1.htm

BMW (2002): *Das Rating-Verfahren nach Basel II und seine Auswirkungen auf die Unternehmen des Automobilhandels.* BMW Financial Services, München.

Borchert J. E. / Goos Ph. / Hagenhoff S. (2004): *Innovationsnetzwerke als Quelle von Wettbewerbsvorteilen.* In: Arbeitsbericht Nr. 11/2004, Institut für Wirtschaftsinformatik, Georg-August-Universität Göttingen

Büschgen, H. / Everling O. (1996): *Handbuch Rating.* Gabler Verlag, Wiesbaden.

Cell Consulting (2004): EU-Osterweiterung bietet Potenziale für Automobilindustrie. In: Excellence Ausgabe II, 2004, Frankfurt am Main

DEKA-Bank (2004): *Emerging Markets: Das Wachstum in den BRIC-Länder ist kein Selbstläufer.* In: Konjunktur, Zinsen, Währungen, Nr. 3, September 2004, S. 8-13.

Delphi (2005): *Warum deutsche Zulieferer den Zug nach Osteuropa nicht Verpassen sollten – und wo die Fallstricke liegen*, Präsentation von Volker Barth, Vizepräsident Delphi und Präsident Delphi Europa, Deutscher Automobil Industrie Gipfel 2005, 14-15. März 2005, Stuttgart

DET-News (2003-09-16.): *Japanese Automakers eye Europe's young drivers.* URL: http://www.detnews.com/2003/autosinsider/0309/17/autos-273010.htm

Deutsche Bank Research (2004-06-16): *Automobilmarkt Osteuropa: Produktionsstandort dauerhaft wichtiger als Absatzmarkt.* In: EU-Monitor vom 16. Juni 2004 S. 11-20.

Deutsche Bank Research (2004-08-10): *Deutschland auf dem Weg zu längeren Arbeitszeiten.* In: Aktuelle Themen vom 10. August 2004.

Deutsche Bank Research (2004-09-17): *Japanische Autos: nachhaltiger Aufschwung erwartet.* In: Aktuelle Themen vom 17. September 2004.

Deutsche Shell GmbH (2001): *Mehr Autos – weniger Verkehr? – Szenarien es Pkw-Bestands und der Neuzulassungen in Deutschland bis zum Jahr 2020.* Shell Pkw-Szenarien, Hamburg.

DIHK (2003): *Produktionsverlagerung als Element der Globalisierungsstrategie von Unternehmen. Ergebnisse einer Unternehmensbefragung.* Deutscher Industrie- und Handelskammertag (DIHK). Mai 2003

Dresdner Bank (2002): *Rating: Ein bewährtes Verfahren gewinnt neue Bedeutung. Die Praxis der Bonitätsbeurteilung für mittlere und große Unternehmen.* Frankfurt am Main.

Dudenhöffer F. / Büttner C. (2002): *Automobil-Standort Deutschland. Teil 1 Empirische Ergebnisse.* Pressekonferenz 8.10.2002, Neue Messe, Leipzig.

Dudenhöffer, F. (2002): *Neue Wachstums-Branche: Die Automobil-Zulieferindustrie* in: Automobile Engineering Partners, Heft 2/2002, S. 4-11.

Dudenhöffer, F. (2003): *Kann Deutschland vom Zulieferer-Wachstum profitieren?* In: Automobile Engineerings Partners Nr. 2/2003, S. 2-6.

DZ-Bank (2004): *Neue Tarifpolitik in der Automobilindustrie – Bremse oder Anstoß für die Konjunktur?* In: DZ-Bank Wirtschaftsbrief Nr. 75 vom 5.11.2004.

Eigermann, J. (2001): *Quantitatives Credit-Rating unter Einbeziehung qualitativer Merkmale: Entwicklung eines Modells zur Ergänzung der Diskriminanzanalyse durch regelbasierte Einbeziehung qualitativer Merkmale.* Sternenfels: Verl. Wiss. und Praxis (= Schriftenreihe Finanzmanagement; Bd. 5).

EIRO (2000): *Outsourcing und Arbeitsbeziehungen in der Automobilindustrie.* European industrial relations observatory on-line. URL: www.eiro.eurofound.eu.int/2000/08/study/tn0008203s.html

Ernst&Young (2003): *Finanzierungsdilemma – Automobilzulieferer: Branche erwartet weitere Fusionswelle.* Ernst&Young AG, Eschborn/Frankfurt a. M.

Ernst&Young (2004a): *Automobilstandort Deutschland in Gefahr?* Ernst&Young AG, Eschborn/Frankfurt a. M.

Ernst&Young (2004b): *Automobilzulieferer: weitere Produktionsverlagerungen nach China und Osteuropa.* Ernst&Young AG, Eschborn/Frankfurt a. M.

FAST-2015 (2004): *„Future Automotive industry Structure (FAST) 2015 – die neue Arbeitsteilung in der Automobilindustrie".* Studie von Mercer Management Consulting und den Fraunhofer-Instituten IPA und IML. In: VDA – Materialien zur Automobilindustrie Nr. 32.

FAZ (2004-08-10): *Neue Runde im Rabatt-Kampf.* In: Frankfurter Allgemeine Zeitung vom 10.08.2004. URL: http://www.faz.net/s/Rub EC1ACFE1EE274C81BCD3621EF555C83C/Doc~E0AF2F9C35B2B4F9DB E0D8E5C29CD7490~ATpl~Ecommon~Scontent.html

FAZ (2004-09-13): *Autohersteller spüren die hohen Rohstoffpreise.* In: Frankfurter Allgemeine Zeitung vom 13.09.2004.

FERI (2002): *FERI Branchen-Rating. Branchenspezifische Kreditrisiken.* In: FERI-Online, Fragen und Antworten, 9.12.2002.

Fraunhofer ISI (2004): *Automobilzulieferer in der Klemme. Vom Spagat zwischen strategischer Ausrichtung und Auslandsorientierung.* In: Mitteilungen aus der Produktionsinnovationserhebung, Nummer 32, März 2004.

FTD (2004-12-02): Reinking, G. *Peugeot schließt Kampfpreis für neuen Kleinwagen aus.* In: Financial Times Deutschland vom 02.12.2004.

FTD (2004-12-05): Reinking, G. *VW rückt von Hochpreisstrategie ab* In: Financial Times Deutschland vom 05.12.2004.

FTD (2004-12-14): Ruch, M. / Reinking, G. *Autokonzerne rücken enger zusammen.* In: Financial Times Deutschland vom 14.12.2004, S. 7.

FTD (2005-02-21): Fromm T. / Reinking, G. *Führungskrise bedroht Fiat-Sanierung.* In: Financial Times Deutschland vom 21.02.2005, S. 8.

Gesamtmetall (2004): *Die deutsche Metall- und Elektro-Industrie in Zahlen ("Zahlenheft").* Gesamtverband der metallindustriellen Arbeitgeberverbände e.V. (Gesamtmetall) URL: http://www.gesamtmetall.de/Gesamtmetall/MEOnline.nsf/id/C35B6E9E2910 2400C1256BB3004E41B0

Handelsblatt (2004-06-29): *Preisdruck zwingt Autozulieferer in engere Kooperationen – Rigides Kostendenken bei den Autokonzernen gefährdet die Struktur der Branche.* In: Handelsblatt Nr. 123 vom 29.06.2004, S. 12.

Handelsblatt (2004-09-02): *Autohersteller drosseln die Fertigung.* In: Handelsblatt vom 02.09.2004, URL: http://www.handelsblatt.com/pshb/fn/relhbi/sfn/cn_artikel_drucken/strucid/P AGE_200012/pageid/PAGE_200038/docid/785564/SH/0/depot/0/index.html

Harbour consulting (2004): *Manufacturing Efficiency – Why is it Important?* Management briefing Traverse city, Presented by: Laurie a. Felax. August 3, 2004.

HAWK 2015 (2003): *HAWK 2015 – Herausforderung Automobile Wertschöpfungskette.* Studie von McKinsey&Company. In: VDA – Materialien zur Automobilindustrie Nr. 30.

Holzapfel, H. / Vahrenkamp, R. (1993): *Fertigungstiefe beeinflusst Verkehr.* In: Logistik Heute, Heft 12, S. 16-17.

IKB (2001): *Automobilzulieferer 2000: Kräftiges Wachstum, differenzierte Ertragsentwicklung.* IKB Deutsche Industriebank, Düsseldorf.

IKB (2003a): *Automobilindustrie – Neue Chancen, zunehmender Investitions- und Finanzierungsbedarf,* Düsseldorf (=IKB-Report – Märkte im Fokus).

IKB (2003b): *Rating für den Mittelstand.* Dezember 2003, IKB Deutsche Industriebank, Düsseldorf (=IKB Information).

IKB (2004): *Automobilzulieferer. Bericht zur Branche.* Dezember 2004, IKB Deutsche Industriebank, Düsseldorf (=IKB Information).

Intra (2004): *Megatrends der Automobilindustrie. Eine Zusammenfassung der wichtigsten Richtungsvorgaben des VDA-Technik Kongresses in Wolfsburg, des Automobilforums in Stuttgart und der IAA in Frankfurt.* URL: www.intra-

ub.de/docs/publikationen/ downloads/ Mega-
trends_der_Automobilindustrie.pdf

iwd (2004a): *Automobilindustrie – Vom Netzwerk profitieren.* In: Informationsdienst des Instituts der deutschen Wirtschaft, Jahrgang 30 Nr. 47, vom
18.11.2004.

iwd (2004b): Schröder Ch. *Die industriellen Arbeitskostender EU-
Beitrittskandidaten.* In: IW-Trends Nr. 1, April 2004.

iwd (2004c): Schröder Ch. *Personalzusatzkosten in der deutschen Wirtschaft.* In:
IW-Trends Nr. 2, April 2004.

iwd (2004d): Schröder Ch. *Produktivität und Lohnstückkosten im internationalen
Vergleich.* In: IW-Trends Nr. 3, April 2004.

iwd (2004e): *Unternehmerische Rahmenbedingungen. Deutschland ohne Lorbee-
ren.* In: Informationsdienst des Instituts der deutschen Wirtschaft, Jahrgang 30
Nr. 51 vom 16. Dezember 2004.

IWK (1999): *Auswirkungen der globalen Marktveränderungen auf die Unterneh-
mensgrößenstruktur in der Automobilzulieferindustrie.* In: VDA (Hrsg.), Materialien zur Automobilindustrie, Nr. 22.

IWK (2001): *Rating als Herausforderung für Mittelstand und Banken – Basel II
und seine Auswirkungen.* Institut für Wirtschaftsanalyse und Kommunikation
Dr. Helmut Becker, München.

IWK (2002): *Automobilindustrie vor der Krise? Entwicklungstrends 2015. Renta-
bilitätskrise der Hersteller. Zulieferer unter Anpassungsdruck.* Institut für
Wirtschaftsanalyse und Kommunikation Dr. Helmut Becker, München.

IWK (2003): *Evaluierung des weltwirtschaftlichen Strukturwandels und der sich
daraus ableitbaren wirtschaftlichen Wachstumspotenziale.* Institut für Wirtschaftsanalyse und Kommunikation Dr. Helmut Becker, München.

IWK (2004a): *Die deutsche Automobilindustrie in der erweiterten EU – Motor der
Integration.* VDA: Frankfurt am Main.

IWK (2004b): *Eroberungsstrategien japanischer Hersteller auf dem europäischen
Automobilmarkt.* Institut für Wirtschaftsanalyse und Kommunikation
Dr. Helmut Becker, München.

Jama (2004a): *Common Challenges, Common Future – Japanese Automakers in
an Enlarged Europe.* Japan Automobile Manufacturers Association, Inc.

Jama (2004b): *The Motor Industry of Japan 2004.* Japan Automobile Manufacturers Association, Inc.

KBA (2005): *Der Fahrzeugbestand am 1. Januar 2005.* In: Pressemitteilung Nr. 5
/2005.

Kurek, R. (2004): *Erfolgsstrategien für Automobilzulieferer – Wirksames Management in einem dynamischen Umfeld*. Berlin et al.: Springer Verlag.

Manager Magazin (2001-10-25): *EURO 500. Methode des Rankings*. URL: http://www.manager-magazin.de/geld/euro500/0,2828,164038,00.html

Manager Magazin (2004-07-23): *Champions League der Konzerne*. URL: http://www.manager-magazin.de/unternehmen/artikel/0,2828,309452,00.html

Mercer (2003): *Studie von Mercer und Fraunhofer-Institut. Die neue Arbeitsteilung in der Automobilindustrie*. Pressemitteilung vom 15.12.2003.

Mercer (2004): *Automobilmarkt China 2010*. Mercer Management Consulting, München.

Moody's (1998): *Ratingmethodologie für Industrieunternehmen*. Sonderbericht. New York et al. Juli 1998

Moody's (2004): *Rating Methodology: Global Auto Industry*. New York et al. September 2004

MWV (2005): *Zusammensetzung des Preises für Superbenzin*. Mineralölwirtschaftsverband e.V., URL: http://www.mwv.de/Grafik_ZusammensSuper.html

OICA (2004): *World motor vehicle production by manufacturer. World ranking 2003*. URL: http://www.oica.net/htdocs/statistics/tableaux2003/Worldranking2003.pdf

OPEC (2004): *OPEC Annual Statistical Bulletin 2003*. Organization of the Petroleum Exporting Countries (OPEC), Vienna, Austria.

Porter, M (1999): *Wettbewerbsstrategie – Methoden zur Analyse von Branchen und Konkurrenten*. Campus Verlag, Frankfurt a. M.

Proies, L. (2004): *Globale Lage der Automobilindustrie: Zukunftschancen und Herausforderungen*. Veranstaltung „Betrieblicher Wandel in der Automobilindustrie und Konsequenzen für die Betriebsratsarbeit" Wuppertal 17.11.2004

PWC (2004a): *Autofacts Executive Perspektives – Quarterly Issue Analysis: Overcapacity*. PriceWaterhouseCoopers URL: http://www.autofacts.com

PWC (2004b): *Gentlemen's dispute or bar room brawl? – Part one: The impact of the new block exemption regulation on carmakers*. PriceWaterhouseCoopers, URL: http://www.pwcglobal.com/ extweb/pwcpublications.nsf/docid/ A545554B87E7DE5780256DB00059BC5B

PWC (2004c): *Quarterly Issue Analysis: Overcapacity*. In: AUTOFACTS. Executive Perspectives. September 2004

PWC (2004d): *Werden sich Angebot und Nachfrage die Waage halten, wenn die Autoindustrie den Osten erobert?* Factsheet. PricewaterhouseCoopers, Automobile Center of Competence.

Radtke, Ph. / Abele, E. / Zielke, A. (McKinsey & Company – Hrsg.) (2004): *Die smarte Revolution in der Automobilindustrie*. 1.Auflage. Frankfurt/Wien: Redline Wirtschaft bei ueberreuter.

S&P (2004): *Industry Report Card: Global Automakers*. Standard and Poors. 06.10.2004. URL:
http://www2.standardandpoors.com/NASApp/cs/ContentServer?pagename=sp /sp_article/ArticleTemplate&c=sp_article&cid=109544106

SAM (2003): *Changing Drivers – Der Einfluss von Klimaschutzstrategien auf Wettbewerb und Shareholder Value in der Automobilindustrie*. SAM & World Resources Institute (WRI), November 2003, Zürich.

Shell (2004): *Shell Pkw-Szenarien bis 2030: Flexibilität bestimmt Motorisierung*. Shell Deutschland Oil, Hamburg.

Spiegel (2005-01-10): Hawranek, D. *Autoindustrie: Fluch der Vielfalt*. In: Spiegel Nr. 2/2005, S. 90-92

Spiegel (2005-02-21): *Arbeitsmarkt – Nivellierung nach unten*, In: Der Spiegel Nr. 8/2005 vom 21.02.2005, S. 82ff.

Spiegel (2005-10-11): Hawranek, D. *Autoindustrie: „Das wird hässlich"*. In: Spiegel Nr. 42/2004, S. 88-90

SZ (2005-01-08): *"Für zwei Jahre alle Regeln aussetzen" – SZ-Interview mit Roland Berger. Wie Deutschland bei der Aufholjagd der Volksrepublik China in der Weltwirtschaft mithalten kann*. In: Süddeutsche Zeitung Nr. 22 vom 08. Januar 2005.

SZ (2005-01-22/23): Reichle J. *Auf ein Neues. Vorschau: Die wichtigsten Autos 2005*. In: Süddeutsche Zeitung Nr. 17 vom 22/23. Januar 2005, S. 17.

SZ (2005-01-28): *Müder Start der Autoindustrie ins neue Jahr*. In: Süddeutsche Zeitung Nr. 22 vom 28. Januar 2005, S. 14.

SZ (2005-03-21): *"2010 sind wir Weltmarktführer". SZ-Interview mit Toyota-Manager Tokuichi Uranishi*. In: Süddeutsche Zeitung Nr. 66 vom 21. März 2005, S. 27.

Tagesspiegel (2004-11-29): *Rackerwochen in den Autohäusern*. In: Tagesspiegel vom 29.11.2004, S. 17.

Toyota (2004): *Toyota in Europe*. Toyota Motor Europe, Brussels, Belgium.

VDA (2001): *Erfolgsstrategien in der mittelständischen Automobilindustrie*. Materialien zur Automobilindustrie Nr. 26, Verband der Automobilindustrie (VDA), Frankfurt am Main.

VDA (2002a): *Allgemeiner Statistischer Informationsdienst – Diskette*. Verband der Automobilindustrie (VDA), Frankfurt am Main.

VDA (2002b): *Auto – Jahresbericht 2002*. Verband der Automobilindustrie (VDA), Frankfurt am Main.

VDA (2002c): *Zukunft des Automobil-Standorts Deutschland*. Materialien zur Automobilindustrie Nr. 27, Verband der Automobilindustrie (VDA), Frankfurt am Main.

VDA (2004-01-29): Das Autojahr *2003: Exportrekord sichert Wachstum – Das Autojahr 2004: Wende im deutschen Automarkt – Initiative für mehr Wertschöpfung in Deutschland*. In: VDA Pressedienst vom 29. Januar 2004. Frankfurt am Main.

VDA (2004a): *Autojahresbericht 2004*. Verband der Automobilindustrie (VDA), Frankfurt am Main.

VDA (2004b): *Die deutsche Automobilindustrie in der erweiterten EU – Motor der Integration*. Verband der Automobilindustrie (VDA), Frankfurt am Main.

VDA (2004c): *Kräftiges Umsatzwachstum in der Zuliefererindustrie – Druck auf Margen bleibt hoch*. In: VDA-Pressedienst vom 24.11.2004, Verband der Automobilindustrie (VDA), Frankfurt am Main.

VDA (2005): *Analysen zur Automobilkonjunktur 2004*. VDA-Jahrespressekonferenz am 27. Januar 2005. Frankfurt am Main.

VDA (2005-01-27): *2004: Hohes Umsatzwachstum – 4.100 neue Arbeitsplätze – Plädoyer für eine "neues Geschäftsmodell Deutschland"*. In: VDA Pressedienst vom 27. Januar 2005. Frankfurt am Main.

Volkswagen (2004): *Automobilindustrie*. Vortrag von Herrn Dr. Uwe Elsner in der VDBE- Arbeitssitzung am 2. Dezember 2004 in Bad Homburg.

Weiss E. (2003): *Branchen-Rating. Zur Erfassung branchenzpezifischer Kreditrisiken*. In: RATINGaktuell, Nr. 2, S. 42-46.

WestLB (2004a): Lier H. / Westin F. *AutoQ. Neither fish nor fowl*. WestLB Equity Markets, Düsseldorf. July 2004

WestLB (2004b): Lier H. / Westin F. *Flexibility as a Competitive Edge*, WestLB Equity Markets, Düsseldorf. June 2004

Wiehle U. et al. (2004): *Kennzahlen für Aktionäre*. Sonderauflage für die Deutsche Post AG. Wiesbaden: comertis.

World Market Research Center (2003): *Japanese Plants lead the way in WMRC European Automobile Productivity Index 2003*. Press release. URL: http://www.wmrc.com/press_release/20030708-1.pdf

ZEW (2004): *Innovationsreport Fahrzeugbau – Ergebnisse der deutschen Innovationserhebung 2003*. ZEW-Branchenreport Innovationen Jahrg. 11 Nr. 11, Zentrum für Europäische Wirtschaftsforschung GmbH (ZEW), Mannheim.

Author

Dr. Helmut Becker, Dipl. Volksw. (graduate economist) and Dipl. Kfm. (graduate in business management), has been managing the *Institute for Economic Analysis and Communication (IWK)*, which he founded, since 1998 which is above all concerned with macroeconomic analyses and forecasts for the preparation of strategic enterprise decisions as well as questions of enterprise communication. A particular focus is the "old economy", especially the enterprises of the automotive industry.

He acquired the necessary knowledge in the course of his long career in science and industry, first of all in the *German Council* ("5 Wise men") and from 1974 as head economist of the BMW Plc. During that time he held numerous posts in the German economy (BDI, VDA etc.).

With the collaboration of:

Yuriy Dutka, Dipl. Volksw. (Graduate economist), born in 1977 in the Ukraine, studied economics at the Ternopil Academy of Economy (Ukraine) and at the University of Passau. Since March 2000 he has been employed by the IWK, first of all as scientific assistant, and in the meantime as Senior Analyst and IT-specialist. His major responsibilities are in the area of empirical economic research as well as analyses of countries and branches.

Niels Straub, Dipl. Volksw. (Graduate economist), born in 1974 in Berlin, studied economics at the LMU Munich. He was first employed as Senior Analyst, after finishing his studies in 2002, with focuses in the area automobile and public banks as well as in the macroeconomic analysis of countries. Presently he is employed as freelancer at the IWK during his postgraduate studies.